Exploring Human Virtue in the Math Classroom

This practical resource guides elementary teachers through ways to ground their mathematics instruction in key human virtues, which enables both educators and students to better connect to the subject matter. As students and their teachers approach math with varying experiences, these common virtues help level the learning field. The chapters are organized around different human virtues – such as beauty, truth, creativity, justice, and love – each providing practical pedagogical examples, teaching strategies to make math come alive, and evidence-based research. Chapters also examine how you can evaluate and modify your existing curriculum and provide reflective prompts to drive conversations with teachers and administrators. Ideal for in-service elementary teachers and as a supplement for pre-service elementary methods instruction, this book empowers you to help your students find deeper meaning in math.

Jonathan Thomas is a Professor of Mathematics Education and Chair of the Department of Teaching and Learning at the University of Kentucky, USA. He is a former U.S. Math Recovery Council board member and chairperson, and mathematics intervention teacher in public, private, and charter schools.

Also Available from Routledge Eye on Education

(www.routledge.com/eyeoneducation)

STEM Matters: Your Guide to Educational Purposes, Policies, Programs, and Practices
Rodger W. Bybee, Patrick L. Brown

The Explore-Before-Explain Guidebook for Science Education: Creating High Quality Lessons for the Classroom and Professional Learning
Patrick L. Brown

Teaching STEAM Through Hands-On Crafts: Real-World Maker Lessons for Grades 3-8
Christine G. Schnittka, Amanda Haynes

STEM by Design: Tools and Strategies to Help Students in Grades 4-8 Solve Real-World Problems
Anne Jolly

Turning it Around: Small Steps or Sweeping Changes to Create the School Your Students Deserve
Todd Whitaker, Courtney Monterecy

How to Get All Teachers to Become Like the Best Teachers
Todd Whitaker

Culturally Responsive and Sustaining Science Teaching: Teacher Research and Investigation from Today's Classrooms
Elaine V. Howes, Jamie Wallace

STEAM Teaching and Learning Through the Arts and Design: A Practical Guide for PK-12 Educators
Debrah C. Sickler-Voigt

Exploring Human Virtue in the Math Classroom

Teaching Practices for Student Flourishing in Grades K-5

Jonathan Thomas

Routledge
Taylor & Francis Group
NEW YORK AND LONDON

Designed cover image: © Getty Images

First published 2026
by Routledge
605 Third Avenue, New York, NY 10158

and by Routledge
4 Park Square, Milton Park, Abingdon, Oxon, OX14 4RN

Routledge is an imprint of the Taylor & Francis Group, an informa business

ISBN: 978-1-032-77788-7 (hbk)
ISBN: 978-1-032-77787-0 (pbk)
ISBN: 978-1-003-48478-3 (ebk)

DOI: 10.4324/9781003484783

Typeset in Palatino
by codeMantra

To the worker bees. To those who put their hands, bodies and minds to work every day, noses to the grindstone, guiding their energies to a brighter future for themselves and all around them, brimming with optimism and hope. And for those who used to feel that hope but lost something along the way.

And mostly, for the incredible children who watch us closely to see what really matters.

Contents

Meet the Author

Jonathan Thomas is a Professor of Mathematics Education and Chairperson of the Department of Teaching and Learning at the University of Kentucky (USA). Jonathan has worked as a mathematics intervention specialist in public, private, and charter schools and is a faculty associate with the Kentucky Center for Mathematics. In that role, Jonathan engages K-12 mathematics teachers to ensure strong connections between research and practice.

Jonathan was born and raised in the small community of Rich Pond, KY (Warren Co.), and served as an infantry paratrooper with the 82nd Airborne Division (U.S. Army) before finding his way to teaching and academia. Even though both of his parents were educators, for much of his childhood and adolescence, Jonathan struggled with mathematics and routinely received low grades in the subject. It was much later that Jonathan came to appreciate how one might experience fulfillment and flourishing in mathematical spaces.

Presently, Jonathan's research is focused on responsive teaching practices in the elementary classroom that help students experience mathematics in a more relevant, impactful, and fulfilling way.

Foreword

I taught math for a number of years before I ever asked a simple question: why is it that for some, math is a hobby – something draws them in so much that they'll do it in their spare time, *for pleasure* – and for others, math is a chore or even a source of deep anxiety?

Most likely, I never thought to ask this question because it hadn't yet occurred to me that every person could truly be influenced in either direction. If you believe, as I once did, that some people are "math people" and others are not, then this question can be easily dismissed, because it would seem patently obvious that only the math people would find pleasure in mathematics. On the other hand, if you embrace the idea – as I now have – that everyone can have a meaningful relationship with math, and that the quality of their experiences around math will shape their enjoyment and affection for it, then this question is more pressing. Why is math a source of joy for some and not for everyone? For the math teacher, it should lead to some soul searching, as it did for me when I finally realized it was an important question.

The conclusion I came to was that we have hobbies (like sport, or music, or craft) because they are life-giving, and they are life-giving because they meet some basic human need that we have. Perhaps you run because you enjoy the beauty of nature or have the desire to grow as a human being. Perhaps you quilt because you delight in being creative and playfully expressing that in patterns. Similarly, those who take pleasure in math have experienced the ways that math can meet basic human desires, for things like community, for beauty, for truth, for freedom and for many more longings central to the human experience. Crucially, because these desires have been met, their experiences of math have been life-giving and have led to their flourishing as human beings. That is why math is a hobby and they keep coming back for more, because it is a source of freedom and delight.

But too often, a person's experiences from their school math days are exactly the opposite. Their agency has been taken away from them. They've been asked to learn procedure after procedure, without the joy of understanding why they work, or the freedom to tinker and to play, to create their own paths to understanding. On top of that, they feel this pressure to do everything *fast*. We have made the math classroom a joyless and lifeless space, and demand that our students become human calculators. No wonder they are bored, or anxious.

The world does not need better human calculators. We already have calculators. What the world needs is people who can think, and reason, who take joy in understanding, who take responsibility for the ways that math is used and misused. In other words, what the world needs is humans who exemplify human virtues. Of course having procedural and arithmetic fluency is important, but it should be secondary to the development of virtue – character qualities that help us flourish and allow us to engage the world in meaningful ways. Such virtues include appreciating beauty, paying attention to wonder, seeking to know truths deeply, delighting in playful thinking, having an eye toward justice, and exemplifying a love for others that believes that everyone is a math person. For the teacher, these aspects of math instruction should remain central and become even more important as computing and AI tools proliferate.

In 2020, I wrote a book called *Mathematics for Human Flourishing*, which makes a case that people need better reasons for learning math, and teachers need better reasons for teaching it. A focus on human flourishing emphasizes that math isn't just instrumental – it isn't just a set of facts you learn for a job or a credential – rather, learning math can actually be *intrinsically good for us*, in building many different virtues and shaping our character and disposition in lasting ways. Being persistent in problem solving is a virtue that will serve you well no matter what you do in life. Having the courage to tackle hard problems is a virtue, borne and built by joyful math experiences, that has similarly lasting effects in every arena of life.

What Dr. Thomas has done in this book is to take the next step – for teachers – in thinking deeply about how a focus on virtue development and human flourishing should impact the practice of a teacher and how a focus on virtue development can bring life to the classroom and make it feel more human. For students lucky enough to have such a teacher, math isn't just something you'll use later (as they're often told), but it is a subject that can feel alive and interesting and joyful *right now.*

You will find Dr. Thomas has many practical suggestions as well as useful pedagogical reflections that will sharpen your own thinking. I also deeply appreciate the ways that he contextualizes his teaching practice to his local community. His examples will likely spur your own creativity. But most importantly, in his many personal stories and reflections, his humanity shines through. In that, he models what math teaching can and should be.

As you grow in your practice of making your math teaching feel more human, may the joy you bring your students be reflected back to you, in wonder and delight.

Francis Su
Benediktsson-Karwa Professor of Mathematics
at Harvey Mudd College
Past President of the Mathematical
Association of America
Author of Mathematics for Human Flourishing (2020), winner of the 2021 Euler Book Prize

1

Introduction

For each chapter, I have curated a playlist from my own music library. These are songs that put me in the mindset of the topics at hand. Some may be on the nose and others may feel like a bit of a stretch. Hopefully, though, you find the music an enjoyable addition to your thinking in these areas.

Note: Spotify automatically adds selections to a given playlist. My original selections are the first 6–8 songs on a given list

https://open.spotify.com/playlist/2WCQDpi2CcSHSbvxTNhT4h?si=dU5X5u80RxmOJvGcJ9ePiw

Tuesday, 11:10am, What's the Point?

It's Tuesday right before lunch and William has his head down on the table again. He was one of my first-grade math intervention students and we had been here many times before. Despite my best efforts to get him to look at me or the math task in front of him, his head remained buried in his arms on top of the table. "C'mon William. Let's give it a go. You've got this". I waited a beat. "Waiting on you. I really need you to be with me here". I waited another 20 seconds or so. William shifted his head so that I can barely see one of his eyes and the corner of his mouth.

DOI: 10.4324/9781003484783-1

"What's the Point?"

It came out as almost a whisper, but it rang in my ears the rest of that morning, the next day, and all the way up to this moment as I write this paragraph. What is the point of teaching and doing math? In the very special way that kids bring us the truth, William brought some truth to me that morning that launched me on my own journey as a teacher and professor. It's not a particularly new or novel question – what's the point? Talking with educational leaders or just a simple Google search (e.g., "why study math?") will provide plenty of plausible answers. Good jobs, getting into college or studying a trade, the importance of foundational math knowledge in the day-to-day of life, and so on. I don't dispute any of these and they are great reasons for giving time and energy to learning math. However, the importance of landing a prized engineering gig or estimating your grocery bill as you throw items into the cart often aren't very compelling to a first grader, and William was no exception. My grown-up arguments for "what's the point" didn't hold much currency in his world. If I'm honest with myself, they don't hold much sway with other adults either, given prevailing social attitudes toward math.

Did you know that an estimated 93% of the population of the United States has some form of math anxiety, which leads to math-avoidant behaviors (Luttenberger, Wimmer, & Paechter, 2018)? Does that estimate surprise you? For a long time in my life, I was buried deep in that 93% of folks who don't understand math, or at least not well, and because of this, tend to avoid math. I got through life just fine. I knew my times tables. I could measure a wall to fit a piece of furniture. That's all I need and don't ask me to engage with this thing called math at more than a superficial level. Please and thank you. But this book isn't about anxiety. I would argue that anxiety is just a symptom of something deeper regarding what we think math is, and consequently, how we teach math in schools. Or, more succinctly, what's the point?

The purpose of this book is to examine how we might connect our math teaching and students' math experiences to deeper and more meaningful human desires (e.g., truth, meaning, freedom, beauty, and love) which are sometimes referred to as virtues.

In his exploration of a mathematics for human flourishing, Su (2020) writes,

> When some people ask, 'When am I ever going to use this?' what they are really asking is 'When am I ever going to value this?' They are equating math's value with utility because they haven't seen that they can value anything more. A grander, more purposeful vision of mathematics would tap into the desires that can entice us to do mathematics as well as the virtues that mathematics can build. (p. 11)

This is precisely the vision that will guide our thinking in the coming chapters, but with a specific eye on specific teaching strategies, rich activities, and culture-setting moves that ground one's math experience in fundamental human virtues.

Let's Play

In the chapters to come, we will explore different virtues and how they relate to math learning in the classroom; however, to illustrate the broader purpose of this book, let's consider one particular virtue, *play*, and how it might shape the experience of students and their teachers. Take a minute or two to consider the following task.

What's the craziest way you can imagine to make 31?

Jot down your ideas here. Share them with your students. See if they can build on your ideas.

____

____

____

____

How wild did you get with that one? If you are like me, you probably started out in a fairly tame place. Well, I can add some things and then subtract some things – say 24 + 52 – 45. That's fine, but maybe not too wild. What other math tools do I have to play with this a bit. What if I wanted to do something with exponents and square roots ($2^5 - \sqrt{1}$). Ok, this is getting a bit more interesting. But what if we hadn't yet learned about those processes and only had four arithmetic operations (+ – × ÷) at our disposal?

$$\frac{94+83+72+61}{4+3+2+1}$$

Admittedly, it took me a few minutes to arrive at this one. My goal was to create a fraction (or, more fundamentally, a division problem) that had a sequential element in both the numerator and the denominator. I experimented with some numbers. I tried several that didn't work. I tried some that did, but I just didn't find the sequences appealing and thought I might create something better. Throughout, I engaged in a spirit of play as I approached this task and I hope you did too. This is a place where the stakes are low. Where "doing it right or wrong" loses importance, and we can escape in our experimentation. In his famous essay, *A Mathematician's Lament,* Lockhart (2002) writes of such mathematical play,

> I'm just *playing*. That's what math is— wondering, playing, amusing yourself with your imagination. So, we get to play and imagine whatever we want and make patterns and ask questions about them. But how do we answer these questions? It's not at all like science. There's no experiment I can do with test tubes and equipment and whatnot that will tell me the truth about a figment of my imagination. The only way to get at the truth about our imaginations is to use our imaginations, and that is hard work. This is what a piece of mathematics looks and feels like . . . asking simple and elegant questions about our imaginary creations, and crafting satisfying and beautiful explanations. There is really nothing else quite like this realm of pure idea; it's fascinating, it's fun, and it's free!
>
> (p. 4)

classroom should invoke some manner of disequilibrium among students. Carter (2008), writes,

> confusion is something you go through, not a permanent state of being. Typically, when students are frequently confused during mathematics class, they assume they are not good at the subject and stop trying . . . [but] if you are not struggling you are not learning.
>
> (p. 136)

The parameters of my task were developmentally on target, but the way that I pitched the task group as a "game" simply wasn't resonating with William – head down, "What's the point?"

So, I phoned a friend. Dr. Les Steffe is a groundbreaking mathematics education researcher who, recently retired, was then at the University of Georgia. Much of what we know about how young children come to develop foundational numeracy and understanding of units comes from his investigations. For a detailed exposition of his pivotal role in mathematics education, I encourage you to read the most excellent commentary on the field that he and his colleague Frank Lester published several years ago (Lester & Steffe, 2013). At that time, though, Dr. Steffe and I were working together on an upcoming workshop focused on early numeracy, and I knew he would have some ideas about how to draw William into a productive math experience aimed at moving away from concrete objects:

> Your kiddos have little toys or trinkets, right? How about something like a memory game, but with objects? Go down to your basement and get three small objects. No, get about five or six. Things that have some visual pop to them and then get a bowl or something to cover them up. See if William can remember, name, and keep track of how many things are under that bowl. Maybe that would work?

Sitting at my kitchen counter, on the phone with Dr. Steffe, I felt the spark of play. "Something like a memory game", he said.

Reflecting on that conversation, I saw another layer of play. Dr. Steffe was playing with his own ideas of mathematics development, hypothesizing an activity that might reach this student in an engaging way, imagining how it might work out, revising his thinking, and so forth. He was playing as a scholar and, over time, I would learn to play with my instruction and math tasks as a teacher and professor.

After I collected a few objects (see Figure 1.2), I met up with William for our next intervention session. While we also worked on numeral identification and some verbal sequencing, my primary activity was to be the game proposed by Dr. Steffe. I had gathered the following six small objects: a yellow chair, blue bear, crystal star, red candy, green frog eraser, and plastic penguin. I also covered a food storage container in blue duct tape to serve as my "bowl".

FIGURE 1.2 Mental Imagery Game Objects

Immediately, William wanted to know what we were going to do with these seemingly random items, and I let him examine and name each one (i.e., "green frog", "blue bear", and "star"). Once we determined there were six objects and had named them, I placed them in a row, asked William to take a mental picture of the arrangement, and then concealed them with the blue bowl. We spent the next several minutes thinking through the varying tasks as I touched the top of the bowl in the approximate location of the items:

> What it on the far left? What's next to that object? How many objects are to the right of the yellow chair? What's on the far right? What's just to the left of that object? Etc.

Periodically, William would need a sneak peak, and after we had mastered one sequence, I would mix up the objects and we would go again. William was all in on this activity.

At first blush, this may not appear mathematical, but again, the goal was to help William develop the mental imagery involved in stepping away from the concrete materials he had relied upon for arithmetic tasks thus far (Steffe, 1992; Thomas & Harkness, 2019; Thomas & Tabor, 2012). From a developmental standpoint, this task was pitched right where it needed to be (*Bravo, Dr. Steffe!*); however, it was more than that. William perceived the task, unlike the covered counters, as an opportunity for play. He would occasionally get the sequence wrong or misname an item, but this did not deter him from trying again in the least. By the end of the lesson, he wanted to play with additional, different items and asked me to bring back more things from my basement. This was my first experience where I consciously considered how the virtue of play can be a profound driver of mathematical activity. Not play from my perspective (e.g., "Let's play a game with these counters"), rather authentic play from the perspective of the students that I sought to engage. What would be truly engaging and perhaps fun from their perspective? It took me many more years before I would explore how math might be connected to or propelled by other virtues.

Virtue and Math

Earlier, in this chapter, I introduced the notion of connecting our math teaching and students' math experiences to deeper and more meaningful human desires, referred to as virtues. William asked, "What's the point?" The answer lies in the profound fulfillment that comes with virtuous activity. Philosophers and scholars have been interested in this sort of fulfillment since the earliest days of recorded thought. Sometimes referred to as *eudemonia*, or living a complete life, the way humans pursue sustained happiness and well-being has proven to be an extraordinarily durable topic of study. In *Nicomachean Ethics*, Aristotle (trans. 2002) distinguished between fleeting pleasures and a deeper sense of well-being derived from "actively pursuing virtues and excellence" (Ryan, Huta, & Deci, 2008, p. 143). How, though, do these lofty ideas connect to the math classroom? In the instance of play, perhaps it is a bit clearer how that virtue might intersect with our teaching. We might consider how our activities and tasks create space for engagement and enjoyment among students as they do some challenging mathematical thinking. However, virtues like truth, beauty, freedom, and justice (among others) might seem more distant from our practice. Love, for example, is almost universally considered virtuous and the source of deep human fulfillment; however, the notion of our math teaching being somehow mediated by or directed toward love may strike you as strange. Certainly, we care for our students deeply, but the concept of love is often perceived as distinct and detached from our daily professional lives of curricula, assessment, learning goals, and the myriad other classroom activities that draw our attention. Yet love, and other virtues are precisely the answer to William's question, "What's the point?" The point of doing anything, including engaging in mathematical activity, is to pursue virtues that lead to a deep and lasting sense of fulfillment, and in the coming chapters, we will examine virtues such as love, truth, beauty, freedom, creativity, and so forth in detail with an eye toward how these virtues might be put into action in our math classrooms.

Here, I offer a quick note on my use of the term, virtue. In many instances, virtue may be considered the human embodiment or enaction of some ideal or goodness. I might read, in an old book, for example, that "The brave knight maintained his virtue despite the temptation of greed" or some such. In this instance, the character embodied the virtue, presumably, of honor. For our explorations, I tend to use the term a bit more fluidly, where virtue serves as a shorthand category for the ideals that, from time immemorial, humans have pursued to find deep fulfillment. In this sense, virtue serves as both an embodiment of an ideal, and an ideal itself. Play, for example, is something we can do and, also, something we can seek. Truth is something to which we can speak and adhere, and, also, something we can pursue. Moreover, these ideals (e.g., play and truth) are fully accessible for us to locate in the mathematics classroom, should we so choose. If that strikes you as a lofty ideal, then I agree. This is a very idealistic book, but idealism is important. As something of a virtue, the pursuit of the ideal (some utopia or perfect truth) while almost never attainable, drives us to improve and become better than we are at present. So, stick with me through these coming chapters and let's dabble in some idealism as we think about how to weave deep and fulfilling virtues into math teaching and learning.

Rehumanizing Math for Students and Teachers

At this point, perhaps you are intrigued but not completely sold on the idea of virtue in the math classroom. That is understandable and I felt that way for a long time, too. The reason, I think, is that we (societally, I mean) see math as something that is separate from the human experience. The math that we encounter is often thought of as an objective reality that exists beyond our minds – the language of the universe so to speak. Bishop (1990) writes,

> the conventional wisdom was that mathematics was culture-free knowledge. After all, the popular argument

> went, two twos are four, a negative number times a negative number gives a positive number . . . Surely, therefore, it follows that mathematics must be free from the influence of any [human] culture?
>
> (p. 51)

While the universality of math makes it a uniquely powerful mechanism to examine all manner of natural phenomena, it always takes place inside of our human minds, with or around other human beings while using human languages. As such, math is a deeply human activity and should be considered as such.

Unfortunately, though, this view of math as neutral, objective, and separate from the human experience often results in what Gutiérrez (2018) refers to as dehumanization. She argues, rather, that as participants in a math community, we should seek to rehumanize math. Gutiérrez writes,

> I use the term *rehumanizing* as opposed to *humanizing* . . . to honor the fact that humans (and other living beings) have been practicing mathematics for centuries in ways that are humane. Among other things, women in India crate elaborate and symmetrical floor patterns (rangoli) with rice that adorn the doorways of homes and get swept away with the daily entry and passage of feet . . . Black women throughout the world create complex curves and spirals through cornrow designs in hair . . . Latinx jazz, mambo, and salsa musicians produce a five-stroke son clave rhythm based on geometry and least common dominators . . . In many ways, we do not need to invent something new; we simply need to return to full presence that which gets erased through the process of schooling . . . When people are encouraged to express themselves through the practice of mathematics, they are more likely to draw upon an innate sense of aesthetics and intuition and to seek ways that are pleasing to them.
>
> (p. 4)

In this passage, Gutiérrez is describing the profound value of repositioning math as a human endeavor – as a powerful tool for human expression, enlightenment, and fulfillment. Moreover, as she writes, this is not the invention of a new idea or process. Rehumanizing means returning to a math that has always been a part of our experience, should we take the time and care to notice. A rehumanized math is one that is deeply connected to virtue and leads to intrinsic fulfillment. In the chapters that follow, we will explore, in very practical terms, the different ways achieve this sort of rehumanized math in our classrooms.

One more thought on rehumanizing math teaching and learning. Thus far, I have centered my arguments for virtue in math classrooms on our student participants. Indeed, connecting virtue to their math experiences is an important way to help them find value and fulfillment within the discipline. Much of our time in this book will focus on students. However, connecting virtue and math in our classrooms is also an incredibly fulfilling way to teach. You may be aware that the United States has been experiencing a teacher shortage for some time and that shortage is only growing more acute. Analysis by the Learning Policy Institute (Sutcher, Darling-Hammond, & Carver-Thomas, 2016) indicates that the United States is experiencing shortages of over 100,000 teachers annually, and this study was conducted several years ago. The American Association of Colleges of Teacher Education (King, 2018) reported that the number of "undergraduate education degrees awarded annually peaked in the 1970s at almost 200,000 and that number is less than 100,000 today" (p. 2). Moreover, in the same report, four years past graduation, only about 60% of education majors remain in the classroom and 16% of education graduates never even set foot in the classroom. While improving conditions, culture, and compensation (Thomas, 2019) all play a part in addressing this growing challenge, rehumanizing the work of teaching, and teaching math for those who do it, is an essential part of finding fulfillment in our professional lives. Indeed, steeping our math classrooms in virtue not only creates a richer and more rewarding experience for our students as well, it can also do the same for us teachers. In the chapters to come, I will explicitly call out moments where virtue-mediated math teaching

makes the work more interesting and enjoyable, but I suspect that as you travel with me on this journey, a connection to professional satisfaction will be self-evident.

Conclusion and Questions for Reflection

For each chapter in this book, I have some questions that hopefully will guide your thinking into new spaces. Connecting virtue and math in our classrooms is complex work, and it is driven in large part by our identities, worldviews, and lived experiences. With that in mind, below are some questions to consider as we conclude this chapter and launch into a world of virtuous mathematics – an exploration I hope you find deeply rewarding and fulfilling:

- *If you had to describe the nature of math a few sentences, what would you say? How would you sell someone on the importance of math? What is the most important thing to know about doing math and why?*
- *Do you enjoy doing mathematics? Do you ever engage in math thinking for fun or look for math in the world around you, or do you tend to shy away from it in your personal spaces?*
- *What is your math biography? How would you describe your relationship to math over time? In what ways has it changed or stayed the same?*
- *When do you most enjoy teaching math and what sorts of things lead to that enjoyment? When do you least enjoy teaching math and why?*
- *How much opportunity do you have to shape the math teaching in your classroom in ways that you think are productive for your students? If you wish that you had more opportunity to direct your math teaching, what is preventing you from doing so? If you do have opportunities to direct your own teaching, do you take full advantage? If so, how? If not, why not?*
- *Imagine you had a math lesson that was entirely of your own making with no concerns for "school constraints". What would you design? What would be your goals or aims? What would you hope that students took from that experience?*

A MATH POEM

A third
And a half of a third
Is a _______

Greenstein, 2023, p. 105

Write your own math poem. It can be as simple or as silly as you like, but make sure it connects to some math idea or element.

____

____

____

____

References

Aristotle. (2002). *Nicomachean ethics* (S. Broadie and C. Rowe, translators and commentary). New York: Oxford.

Bishop, A. (1990). Western mathematics: The secret weapon of cultural imperialism. *Race and Class, 32*, 51–65.

Carter, S. (2008). Disequilibrium and questioning in the primary classroom: Establishing routines that help students learn. *Teaching Children Mathematics, 15*, 134–137.

Gutiérrez, R. (2018). The need to rehumanize mathematics. In I. Goffney, R. Gutiérrez, & M. Boston (Eds.), *Rehumanizing mathematics for Black, Indigenous, and Latinx students* (pp. 1–10). Reston, VA: National Council of Teachers of Mathematics.

King, J. E. (2018). *Colleges of education: A national portrait.* Washington, DC: American Association of Colleges for Teacher Education.

Lester, F. K., & Steffe, L. P. (2013). Establishing mathematics education as an academic field: A constructive odyssey. *Journal for Research in Mathematics Education, 44*, 353–370.

Lockhart, P. (2009). *A mathematician's lament: How school cheats us out of our most fascinating and imaginative art form.* New York: Bellevue literary press.

Luttenberger, S., Wimmer, S., & Paechter, M. (2018). Spotlight on math anxiety. *Psychology Research and Behavior Management, 11*, 311–322.

Ryan, R. M., Huta, V., & Deci, E. L. (2008). Living well: A self-determination theory perspective on eudaimonia. *Journal of Happiness Studies, 9,* 139–170.

Steffe, L. (1992). Learning stages in the construction of the number sequence. In J. Bideaud, C. Meljac, & J. Fischer (Eds.), *Pathways to number: Children's developing numerical abilities* (pp. 83–88). Hillsdale: Lawrence Erlbaum.

Su, F. (2020). *Mathematics for human flourishing.* New Haven, CT: Yale University Press.

Sutcher, L., Darling-Hammond, L., & Carver-Thomas, D. (2016). *A coming crisis in teaching? Teacher supply, demand, and shortages in the U.S.* Palo Alto, CA: Learning Policy Institute.

Thomas, J. (2019). The big short. *School Science and Mathematics, 119,* 429–431.

Thomas, J., & Harkness, S. S. (2019). Tethering towards number: Coordinating cognitive variability and stage-oriented development in children's arithmetic thinking. *Mathematics Education Research Journal, 31,* 325–347.

Thomas, J., & Tabor, P. D. (2012). Developing quantitative mental imagery. *Teaching Children Mathematics, 19,* 174–183.

2

Meaning and Perseverance

For each chapter, I have curated a playlist from my own music library. These are songs that put me in the mindset of the topics at hand. Some may be on the nose and others may feel like a bit of a stretch. Hopefully, though, you find the music an enjoyable addition to your thinking in these areas.

Note: Spotify automatically adds selections to a given playlist. My original selections are the first 6–8 songs on a given list

https://open.spotify.com/playlist/6XtausWJaECHjnM1fSusrj?si=dwekSCARS6SC4gJQg3LrMA

Malik is So Done with This One

A while back, a friend of mine, who teaches second grade, shared some video from her classroom with me. Let's call her, Britt. We had encouraged one another to video record our teaching and share it with each other. In this video, she captured one of her students, let's call him Malik, working on the task 400–198. This task has particular significance in portions of the mathematics community (intervention, in particular) due to the way it allows students to experience and explore the limits of the standard algorithm for subtraction (MacDonald & Thomas, 2023; Wright, Martland, & Stafford, 2006). Take a moment and

DOI: 10.4324/9781003484783-2

imagine "stacking up" 400 and 198, and then going through the procedural machinations of subtraction in that algorithm. All of that borrowing and carrying sounds, to me at least, like a real chore. However, if I had come to understand multi-digit subtraction primarily using the standard algorithm (as many students do), contemplating alternate ways of solving this may appear daunting – those numbers are pretty big, right?

Returning to Malik, in the recording, Britt approaches him while he is staring at the task 400-198 on his paper. She had asked her students to try and devise another way to solve this problem besides the standard algorithm and be prepared to talk through their strategies. Below is a brief exchange from that video:

BRITT (TEACHER):	"So, 400 minus 198. What can we do with that?"
MALIK (STUDENT):	"Are you sure we can't use paper?"
BRITT:	"Well, you can use paper if you need to, but see if you can come up with a strategy that's different from stacking them up and subtracting"
MALIK:	"I mean, I can try to think about it".
BRITT:	"Ok, I'm gonna hang here for a second while you think. When you get done thinking, tell me what you were doing".
MALIK:	"*Long pause of about 15 seconds* . . . I don't know. Can I do another one? I am so done with this one".
BRITT:	"Sure . . . *pauses to think* . . .How about 100–97?" [writes 100–97 near bottom of Malik's paper]
MALIK:	"That one's easy. It's 3".
BRITT:	"How do you know it's 3?"
MALIK:	"Just count up to 100 . . . from 97, you know".
BRITT:	"Gotcha . . . *pauses to think* . . . What about 300–196?" [writes 300–196 on Malik's paper]
MALIK:	"*thinking for about 10 seconds* . . . 104! It's the same thing. Just count up from 196 and then you have to go up 100".
BRITT:	"Nice strategy. So you move up, I mean count up to the next 100 and then add another 100. How would you. . ."

MALIK:	"*interrupts abruptly . . .* Give me the 400–198 one".
BRITT:	"OK, what do you have for that one?"
MALIK:	"It's 202. It's actually really easy. You just count up to 200 and then it's another 200 that gets you to 400".
BRITT:	"Nice! I like it! We are going to share out in a few minutes. See if you can come up with a good way to describe your strategy to the class. Try to come up with a name for it, too".

Thinking about this exchange, there are many places we might engage; however, there are a few elements that are particularly noteworthy to our discussion of meaning and perseverance. At the onset, Malik appears unable to engage with the task 400–198. In a very real sense, in this moment, Malik's meaning of the task is constrained such that he prefers to move on to something else ("I am so done with this one".). Britt then responds with a series of tasks (100–97; 300–196) that create space for Malik to reason mathematically (apart from the standard algorithm). The first one (100–97) seems pretty easy for him, but the second one (300–196) requires a bit more thinking. Or, in other words, Malik has to persevere with his problem solving. When successful, he then continues to persevere by returning to the original task, 400-198, that he initially found so daunting. As Britt and I watched this moment unfold on the recording, we realized we were observing a real-time change both in the way that Malik was making meaning of a mathematical operation, as well as his willingness to persevere in that pursuit of meaning. These two ideas (or virtues, if you will), *meaning* and *perseverance,* will be our landscape for exploration in this chapter.

The Virtue of Meaning

It's hard to pick a particular virtue that rises above others when it comes to teaching and learning mathematics, but *meaning* would be a contender for the prize. After all, a thing must have meaning to an individual before one can imbue or ascribe to it love, creativity or any number of other virtues. On this point, Su (2020) describes how meaning may unlock contemplation and perseverance, the latter of which happens to be a key topic of this chapter. Indeed, things that we find inscrutable and meaningless tend to evaporate in our minds very rapidly. We may be mandated to engage with meaningless things (in a mathematics classroom, for example), but there is a reason we have the saying *in one ear and out the other.* The reason ideas, concepts, and processes tend to quickly travel the width of our heads before exiting (metaphorically speaking) is that we never found meaning in them. Great thinkers of antiquity gave considerable attention to the virtue of meaning. Aristotle (1925-trans.) writes in the seminal work, *Metaphysics*, "all [individuals] by nature desire to know" and this search for meaning has driven much human activity for much of human history.

Moving forward in time, an esteemed scholar of mathematics education, William Brownell (1947), writes on the topic of meaning:

> We should . . . distinguish what I shall designate the *meaning of* a thing and the *meaning of* a thing *for* something else; for the sake of brevity, between *meaning of* and *meaning for.* I know little about the meaning *of* the atomic bomb because I lack the knowledge of chemistry and physics which are requisite to accurate understanding, but I think I know a good deal about the meaning of the atomic bomb *for* other things – for peace or for the destruction of our culture, for example . . . that children have meaningful experiences when they use arithmetic in connection with real life needs, relates to meanings *for* . . . On the other hand, just as the meaning *of* the atomic bomb is to be found in the related physical sciences, so the meanings *of* arithmetic are to be found in mathematics . . . They must be sought in the mathematical relationships of the subject itself, in its concepts, generalizations, and principles. In this sense a child has a meaningful arithmetic experience when the situation he deals with "makes sense" mathematically. He behaves

> meaningfully with respect to a quantitative situation when he knows what to do arithmetically and when he knows how to do it [*meaning for*]; and he possesses arithmetical meanings when he understands arithmetic as mathematics [*meaning of*] . . . Meaningful arithmetic, in contrast to meaningless arithmetic, refers to instruction which is deliberately planned to teach arithmetical meanings and to make arithmetic sensible to children through its mathematical relationships.
>
> (pp. 256–257)

In this passage, Brownell is describing the different forms of meaning that one might construct. In mathematics, one might develop a meaning of a particular concept or process. For example, a first-grade student might realize that, when given the task 12 + 4, she is able to count on from composite 12 rather than start at one and count all the way to 16. In this instance, the student is working in the *meaning of* space. In this sense, she has achieved a key insight (or meaning) into the mechanics of units and arithmetic operations. However, if this student then realizes that this is a much more efficient way to keep track of collections her toys, then she is working in the *meaning for* space. In this instance, the student has ascribed some meaning or purpose for this arithmetic activity such that it has an understood value in her lived experiences. Math teachers sometimes refer to these ideas as conceptual understanding (*meaning of*) and relevance (*meaning for*), but foundational to all of this is the importance of meaning.

One complicating element of math learning deals with how ideas and processes move toward the abstract. For example, students may first learn about fractional quantities by dividing whole candy bars onto portions, then move toward representations of those activities in the form of bar models, and ultimately to purely symbolic expressions of operations involving fractions. Webb, Boswinkel, and Dekker (2008) describe this movement from informal (candy bar squares) to preformal (bar models) to formal (numeric symbols) as the *progressive formalization* of representations and notations. Fundamentally, in the ways that individuals develop math knowledge, skills, and practices, movement is toward a more decontextualized abstract and this direction would seem to undermine our search for meaning. On this complication, Su (2020) writes:

> People often think of abstraction as stripping away meaning. But in fact, abstraction does the opposite – it enriches meaning. When you see that two things have similar structures or behavior, then those similarities create a connection, a new meaning for you that wasn't there before . . . if only you've ever seen one dog, you might think that a dog must be a German shepherd. Once you've seen several, you begin to realize that the meaning of *dog* is richer than you realized. Abstraction enriches meaning by helping you take a collection of examples and to see just what is essential about, for instance, the "dogness". In so doing, you see what's the same about many different things.
>
> (p. 45)

Indeed, it is that richness of abstraction in mathematics that can propel meaning across settings and contexts. However, there is not one *right path* or teaching strategy that leads us in this direction. Rather, we might think of fostering meaning as a broader goal that carries with it many different instructional practices depending upon the students and the mathematics at hand. One essential feature of constructing meaning is that it often does not happen without perseverance.

The Virtue of Perseverance

In the earlier example involving Malik and Britt, there is a moment where Malik pauses for ten seconds while contemplating the task 300-196. Likely, he is trying to make some connection to the task immediately preceding (100-97) that was quite easy for him. While ten seconds is not an inordinate amount of time, it is the internal processing space Malik needed to think about how he might approach the task. Or, from another perspective, it was the space where Malik elected to persevere with his mathematical thinking. In a most general sense, we can think of perseverance as continuing on when faced with some challenge or difficulty. In teaching and learning of mathematics, though, we may consider this virtue through an intellectual lens. King (2014) writes the following on intellectual perseverance in discussing the virtuous tenacity of Helen Keller and Tycho Brahe,

> We see agents involved in paradigmatically intellectual activities – language learning, public dissemination of ideas, and scientific research. These agents display characters of personal excellence in their pursuit . . . Accordingly, in their possession and exercise of intellectual perseverance, these agents display intellectual virtue.
> (p. 3506)

In this article, King leverages Aristotelean arguments to describe how perseverance of the mind (continuing in the face of intellectual challenge) is a virtue worthy of pursuit and one which contributes to flourishing. On this point, Su (2020) writes the following regarding the valuable nature of persistence and struggle:

> The struggle that I want to amplify as a basic human desire is the struggle to achieve internal goods – perhaps more succinctly described as the *struggle to grow* . . . The struggle to grow is an attraction of the mathematics experience. Math explorers relish interesting puzzles and hard problems. We know what it's like to struggle with a problem for a long time and possibly get nowhere with it. We learn to *enjoy* the struggle . . . In the math education community, the term *productive struggle* describes the state of actively wrestling with a problem, persistently trying out various strategies, taking risks, being unafraid of mistakes, and progressing incrementally in understanding the underlying ideas. This wrestling produces a certain kind of *endurance* which enables us to be comfortable with the struggle.
> (pp. 119–120)

In this passage, Su portrays perseverance as a healthy embrace of struggle which serves as a productive and even enjoyable channel for one's intellectual energy. This last bit, enjoyment of struggle, may strike you as a bit far-fetched, but anyone who has ever watched a young person play a challenging video game (or played one yourself) knows that, in the right circumstances, perseverance in the face of difficulty can be immensely engaging.

Building a Classroom Culture of Meaning and Perseverance

Assuming you are warming to the inherent human goodness of meaning and perseverance, it is fair to wonder how we might foster these virtues in our classroom. Toward this end, one of our first goals as instructional leaders should be to build a classroom culture that creates space for perseverance in the pursuit of meaning. Carter (2008) provides a beautiful account of her creation of a student community that sought meaning and persevered in the face of struggle, which she terms *disequilibrium,* and how such struggle became an opportunity rather than a barrier. Carter writes,

> [When most of the class signaled that they were confused or stuck], I stopped the lesson to discuss disequilibrium. 'This is the feeling of disequilibrium' I explained to my first graders, 'and now the real learning can begin'. Our motto became, 'If you are not struggling, you are not learning' . . . Integrating the term *disequilibrium* into our mathematics class eventually led to a discussion of what to do when you become confused about a mathematical concept. How do you get past your disequilibrium to achieve understanding? Early in the school year, I had introduced the idea that telling someone the answer does not help them learn; in fact, it stops their learning. We had established classroom routines whereby students helped each other by asking questions to move the thinking along without telling the answer or the 'right way' to do it. Now the climate in our classroom relaxed as students recognized that their questions held more value than their answers.
>
> (pp. 136–137)

In this passage, Carter describes how she worked to reshape students' attitudes toward struggle – that intellectual struggle and perseverance were a virtue (i.e., "If you are not struggling, you are not learning") and something to be embraced rather than something to be feared and avoided. In Carter's classroom, the journey toward mathematical perseverance was central to their classroom culture. Moreover, the combination of these virtuous pursuits (*meaning* and *perseverance*) as core cultural element of a classroom is a particularly potent combination as students lean into struggle and disequilibrium as a necessary part of their progress toward constructing a deeper meaning of a mathematical idea. This goal is affirmed by the authors of the Common Core State Standards for Mathematics (NGA & CCSSO, 2010) in their Standard for Mathematics Practice (SMP) 1, *Make sense of problems and persevere in solving them.* One additional element of Carter's example worth noting is that her work of culture building took time. She began early in the school year with her messaging around meaning, struggle, and perseverance, and she continued to lift those ideas throughout the school year. Building culture is not a one-time affair, and requires a committed effort to revisit, reiterate, and generate buy-in from students – and this is true for our culture shaping activities around all the virtues examined in this book. Provided we remain committed to culture building in the spirit of Carter's example, we can then focus on building a particular core practice to help our students seek meaning and persist in that journey.

Core Practice: What's the Problem?

Arguably, when thinking of meaning and perseverance, one core practice that's worthy of our energy relates to the creation and/or selection of the math problems themselves. Without rich and engaging mathematical tasks, students have no opportunity to seek some deeper meaning nor the impetus to persist in some intellectual struggle. Toward this end, a good initial goal is to frequently provide students with tasks that are non-routine. Mayer and Hegarty (1996) write,

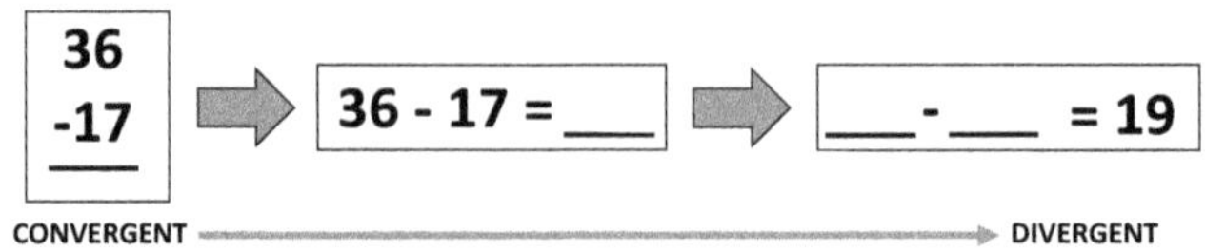

FIGURE 2.1 Convergent and Divergent Tasks

> routine problems are not really problems at all because the problem solver knows what to do and how to do it. For this reason, such problems are often referred to as *exercises*. A non-routine problem exists when a problem solver has a problem but does not immediately see how to solve it.
>
> (p. 32)

It seems quite simple on the surface, but we must ensure that the mathematical tasks, or problems, in which students engage are actually problems from the perspective of the student.

A good test regarding the extent to which a task is routine or non-routine is how much the task converges or diverges from a singular "right answer" or "right way to solve". More divergent tasks tend to be more open-ended. Perhaps they have multiple "right answers" and/or they do not signal to the student exactly how the problem is to be solved (see Figure 2.1).

In this example, we can see how an arithmetic task is at its most convergent when, not only is there a single correct answer, but the vertical task presentation itself is highly suggestive of the standard algorithm. More divergent is the horizontal presentation that is less suggestive of a strategy, and even more divergent is a related arithmetic task where there are many correct answers. The key is not to think of convergent/divergent (and routine/non-routine) tasks as a light-switch that turns on and off, but rather more like a dimmer switch where we can adjust the extent to which tasks are more open-ended along a continuum. At their most open, mathematics presents an opportunity to notice and wonder. In that spirit, consider the equations below. *What do you notice? What do you wonder?* (Figure 2.2).

Perhaps you noticed that all the products end in 5, *or that they all end in 25.* Maybe you began to wonder if this is true for other

$$
\begin{aligned}
5^2 &= 25\\
15^2 &= 225\\
25^2 &= 625\\
35^2 &= 1225\\
45^2 &= 2025
\end{aligned}
$$

FIGURE 2.2 Equations involving Exponents

$$
\begin{aligned}
5^2 &= 25\\
15^2 &= 225\\
25^2 &= 625\\
35^2 &= 1225\\
45^2 &= 2025
\end{aligned}
$$

TEACHER: "*Think about this for a few minutes. What jumps out at you? What are some things you notice? What do you start to wonder? What connections can you make?*
KAYDEN: *"There are a lot of fives"*
MYLES: *"The exponent gets bigger by ten"*
TEACHER: *"The exponent or the base, Myles?"*
MYLES: *"Yeah, the exponent, no the base."*
AMARI: *"All of the products? Those are products, right? They end in twenty-five"*
NAVEAH: *"They grow in, like, a pattern. Twenty-five is like, zero, two, five. Then it's two, two, five, then it's six, two, five. So, zero, two, six, twelve, twenty. It's like two, four, six, eight, and the next should be ten"*
TEACHER: *"So what's going on here?"*
AMARI: [solves 55^2 on calculator] *"So fifty-five is 3025, which ends in 25*
NAVEAH: "*Yeah, which keeps the pattern, up ten from 2025*
KADEN: [solves 26^2 and 46^2 on calculator] *"I checked twenty-six and thirty-six and they both end in six. It's 676 and 1296"*
AMARI: *"But that's not the pattern for fives. Those should end in thirty-six, right?"*
MYLES: [solves 44^2 and 54^2 on calculator] *"The fours end in six, too. Like 'something-four' squared, ends in six. Weird."*
TEACHER: *"So let's stick with the fives for a minute. What could you do to dig into that pattern? How could you break it down? What do you know about exponents or multiplication even to get inside that one?*

FIGURE 2.3 Students Seeking Meaning

exponents – if $7^2 = 49$, then does 17^2, 27^2 and others similar all end in 49? After some exploration, you might have discovered that they (e.g., 17^2, 27^2) don't end in 49, but they do end in 9. What about 4? Do 14^2, 24^2 and so on all end in 16? If not, why not? Is there something special about 5? Mathematics, including those topics in our standards and curricula, is ripe with opportunities for the pursuit of meaning provided we open the playing field a bit. Consider how students engage with this task (see Figure 2.3).

In this example, we see students engaging in a non-routine task – a task that is divergent in that there are many different productive mathematical avenues students could travel following the teacher's prompts. What are some things you notice? What do you start to wonder? In this instance, the students detect several different patterns (e.g., ending in 25) and begin to probe how other exponents may or may not follow a similar pattern (e.g., 26 and 36; 44 and 54). Note, the students' use of calculators comports with SMP 5 – use appropriate tools strategically (NGA & CCSSO, 2010). In this instance, the focus of the task is not calculation, but rather exploration and calculators are a wonderful tool to aid the students' investigation. At the end of the exchange, the teacher recenters the students on "the fives" pattern and tasks them with diving deeper into the structures of that sequence. The students pair up and decide to use what they know about representing partial products in their investigation (see Figures 2.4 and 2.5)

In these examples, students are persevering in substantive mathematical activity to construct a deeper meaning of the patterns

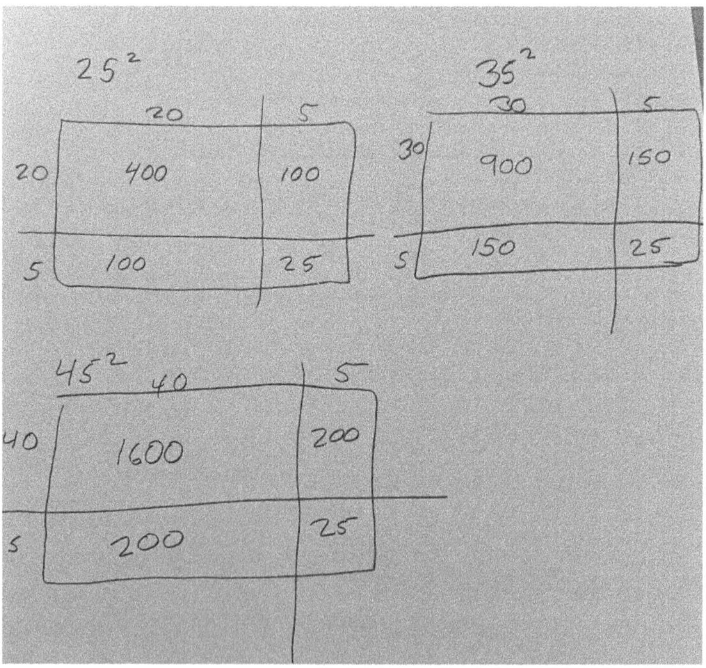

FIGURE 2.4 Amari and Myles Partial Product Recording [Recreated]

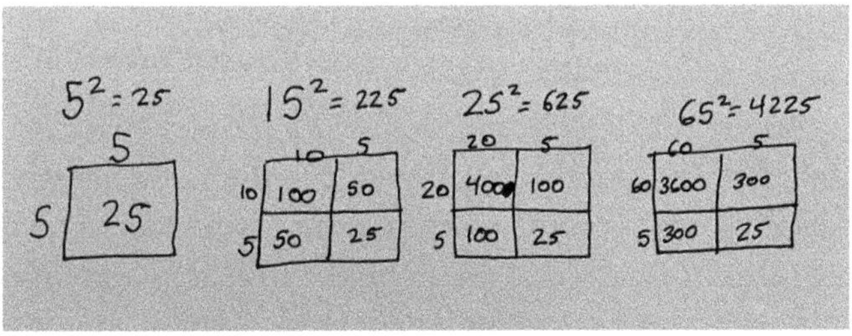

FIGURE 2.5 Kayden and Neveah Partial Product Recording [Recreated]

they have observed. Portrayals of partial products, in particular, helps illustrate why the "fives pattern" (5^2; 15^2 25^2; etc.) all end in 25. For each instance, the "middle products" add to an even hundred which leaves the "final product" of 25 as the concluding element. Importantly, this observation can lead to further exploration – does this structure hold into three-digit numbers or into four-digit numbers? Fun fact, $275^2 = 75{,}625$. What about partial product examinations of 6^2; 16^2 26^2? This type of exploration that rich, divergent tasks provide is highly productive. Lannin et al. (2013) presented nine characteristics of effective mathematics problems and arguably this task is consistent with eight of these criteria including,

> [1] aligns with relevant mathematics content standards, [2] encourages the use of multiple representations, [3] provides opportunities for students to develop and demonstrate the mathematical practices, [4] involves students in an inquiry-oriented or exploratory approach, [5] allows entry to the mathematics at a low level (all students can begin the task) but also has a high ceiling (some students can extend the activity to higher-level activities), [6] connects previous knowledge to new learning, [7] allows for multiple solution approaches and strategies, [and] [8] engages students in explaining the meaning of the result.
> (p. 8)

More broadly, though, is that this task is an example of mathematical space that allows for the seeking of meaning through perseverance and the tasks that we create or select serve as the core ingredient to that process.

Connecting Practice of Professional Noticing

For each chapter in this book, we will examine one *core practice* that connects to the virtues in play. While there are certainly other instructional practices that we may connect to meaning and perseverance, arguably, the mathematical tasks and activities in our classrooms are at or near the center of these virtues. There is another instructional practice, though, that is worth examining throughout this book. That practice, professional noticing of children's mathematical thinking (referred to as professional noticing hereafter) is a key element of empowering mathematics teaching that we may connect to each virtue – a connecting practice, if you will.

Professional noticing is pretty much what it sounds like. It is the paying of close attention to our students, thinking hard about what we see them doing and how they appear to be thinking, and then devising some productive next step. In academic terms, we may consider professional noticing to have three components (Jacobs, Lamb, & Philipp, 2010; Thomas, 2017). *Attending* describes our deliberate focus on the details of a mathematical moment and what our students are doing and saying in that moment. *Interpreting* refers the way that we use those details to construct some understanding of that moment. For example, if I attend to a first-grade student physically touching and moving blocks to count them, I may interpret that behavior to mean that the student relies on physical materials to determine the numerosity of a collection. If I really want to get fancy with my interpreting, I might determine that the child is currently constructing and working with perceptual unit items in numerical contexts (Steffe, 1992; Thomas & Tabor, 2012). Lastly, *Deciding*, involves the devising and enacting some productive next step based on one's interpretation of the mathematical moment. In this example, that might be to present the student with an arithmetic task involving screened or concealed materials. Imagine that, perhaps later in the lesson, when working with this student one-on-one, (or in a small group of conceptually similar peers), I *decide* to pose the following task (see Figure 2.6). "There are four penguins standing there, and three more jumped into the cup. How many penguins are there altogether?"

FIGURE 2.6 Partially Screened Arithmetic with Penguins and a Cup

The purpose of this decision is to present the student with an opportunity to construct imaginary units and move beyond the use of physical items in mathematical settings. Or, more succinctly, can this student count and keep track of penguins that she can't see (in the cup) by imagining them in her mind? Key to this practice, though, is the strong connective thread between attending (seeing the student using blocks), interpreting (determining some reliance on physical materials), and deciding (creating an opportunity to construct a more sophisticated type of unit).

At its core, professional noticing involves watching (and listening to) our students closely, thinking about what this means with respect to students' understanding, and then acting upon that interpretation in some useful way. If that description sounds, to you, like just fundamentally good teaching, then you are not alone. When I first encountered professional noticing, I thought to myself, of course, that's just what good teachers do. We watch our students, think about what it means, and then take action. That's just good teaching, right? Well, in one sense, of course it is; however, we don't develop sound professional noticing through experience alone. Researchers who studied professional noticing among preservice teachers, new teachers, and veteran teachers found that there was no correlation between good professional noticing and years of experience. Rather, the teachers who were good at professional noticing were the ones who intentionally practiced such noticing with an eye toward connecting the components. These teachers mindfully considered the details of a moment, consciously used those details to construct their interpretation, and then, based on that interpretation, devised some next instructional move. To these teachers, there was a clear connection between what they were seeing/hearing, what they were thinking, and what they were doing.

If I am honest, I have not always been the best at this. Sometimes, I would jump to an interpretation. For example, I might say to another teacher, "Janna and Mariah have really strong foundational numeracy". However, if that teacher then asked me, "What did you see that leads you to that interpretation?", I have not always been able to connect that back to some specific observations of those students. Rather, I was operating more on some broader perception or perhaps even intuition. More common, for me, was jumping to a decision. In my quest to find rich and engaging mathematical tasks and activities, has someone asked me why I made those decisions, I am not sure I could trace that back to an interpretation, which itself, was grounded in some attention to students' mathematical activity. There have been many moments, particularly early in my teaching career, when I saw something online or at a conference, I exclaimed, "What a great activity! I am definitely going to do that in class on Monday". In these moments, I was jumping straight to the decision. However, the key to professional noticing is to begin with what we see and hear from our students in the classroom and let the interpretations and decisions flow naturally and intentionally from their actions and interactions.

Professional Noticing for Meaning and Perseverance

Professional noticing in the service of meaning and perseverance takes on some special characteristics. While it may be theoretically impossible to avoid engaging in a form of such noticing (provided we are conscious and engaged as teachers), focusing our practice this way involves some specific considerations. Beginning with attending, we want to watch and listen to students closely, but with particular emphasis on how they are engaging with the task. We might notice how quickly they engage and move through an activity as a signal of difficulty (or lack thereof). We might attend to body language (e.g., head down on table) or words (such as Malik's "I am so done with this one") as indicators that the task is a bit too challenging. We look closely at the strategies students are bringing to bear on a given task. For example, is a student engaging in long counts-by-ones to solve a multi-digit arithmetic task? In these moments, we are trying to absorb as much detail from our students' activities as possible to bring to the next phase – interpreting.

When interpreting our students' mathematics, we might consider how what we are seeing is creating or constraining opportunities for meaning and perseverance. Returning to Carter (2008) and the notion of disequilibrium (or healthy struggle), we want to consider the extent to which a task matches with the student's current mathematical understanding. Indeed, the goal is for the demand of a task to be right at the limit of a child's capability such that there is opportunity to experience and navigate disequilibrium. Vygotsky (1978) used the term *Zone of Proximal Development* (or ZPD) to describe the space between what one can do independently and with some form of assistance or support. Colloquially, ZPD is often used to describe the Goldilocks-edge of one's understanding and where a task might optimally be deployed so that the experience is neither too hard or too easy. These are the spaces that support perseverance, and as we interpret our students' mathematics, this is the lens through which we look. Simultaneously, though, this sort of interpreting requires teachers to know something about the mathematical landscape itself and how concepts develop over time. Clements and Sarama (2009) refer to this sort of conceptual progression as a learning trajectory in that they describe the pathway of how knowledge is constructed by students. Returning to the earlier example with the penguins (see Figure 2.6), the interpretation of the students' physical interaction with tools and how that activity connotes a certain type of conceptual understanding (perceptual units, in this case) is based on a teachers' knowledge of a foundational numeracy learning trajectory – how units move from physical to imaginary and on to more abstract composite structures (Steffe, 1992). While we, as teachers, may not have the comprehensive knowledge of researchers regarding such learning trajectories, we must know, in general terms, the developmental pathways for math concepts that we teach. We need to be able to interpret a child's activities and consider where that individual resides on a given mathematical path and what comes before and after. That way, if we ascertain that a student's struggle is too profound, we can adjust the task or provide a support to make it more accessible. Alternately, if the task is too easy, we can advance it such that it falls more squarely in the student's ZPD. Key, in these moments, is to be attending to and interpreting students' mathematical engagement with a mindfulness of disequilibrium, productive struggle, and opportunities for sense-making.

Lastly, when using our interpretations to make decisions, it is important to ensure that we maintain space for perseverance and meaning making. Rather than trying to pass along our meaning, productive decisions should aim toward creating opportunities for students to construct their own meaning and persevere in doing so. This might involve adjusting tasks upward or downward in difficulty, providing or removing supports and scaffolds to optimize access and struggle, or in some cases, abandoning activities entirely and beginning anew with something else. Other times, our decisions might take on the tone of inquiry through the use of questioning students. Such questions might be diagnostic in nature as we seek additional information about how students are thinking and acting in a given moment. Other times, questioning might be aimed at helping students consider and reflect on key ideas. Whether our decisions center more on tasks or questioning, the key to deciding for meaning and perseverance is to avoid stepping into the moment and doing too much of the thinking on behalf of the student. In a classic practitioner article titled "Never Say Anything a Kid Can Say", Reinhart (2000), a middle grades math teacher, provides some practical tips for these sorts of decisions.

For example, when students so frequently ask, "is this the right answer", Reinhart writes, "my usual response to this question might be that I'm not sure, can you explain your thinking to me. As soon as I tell a student that the answer is correct the thinking stops" (p. 482). One of my favorite tips, though, is to never carry a pencil or pen. Reinhart writes, "If I carry a pencil with me, I am tempted to do the work for the student. Instead, I must ask thought-provoking questions that will lead to understanding" (p. 483). Guilty as charged, Mr. Reinhart, for I have often reached over a student's shoulder (with pencil in hand) with the intent of providing some minimal assistance but wound up doing the majority of the task while the student merely watched. In those moments, I foreclosed on that student's opportunity to think and act mathematically. Rather, deciding in its most productive form keeps the mathematical thinking squarely on the student and maximizes the mathematical opportunity for perseverance in the service of meaning.

Questions for Reflection

At the end of each chapter, I will present some questions that, hopefully, drive some reflection, introspection, and perhaps conversation among you and your colleagues. Given the richness and complexity of the ideas in play, we need some space to consider what this might mean for our unique contexts, how we might take these ideas and run with them, and what challenges we might face along the way. These questions do not have easy answers, but my hope is that they propel you in a productive direction and keep you thinking hard about meaning and perseverance in the math classroom.

- *How well do the tasks and activities in your classroom foster deep mathematical meaning among your students?*
- *How well does your school or district curriculum foster such meaning? What do you do when you feel that your curriculum or learning resources do not sufficiently address mathematical meaning regarding a particular concept?*

- *How do you know when your students have constructed meaning and understand a math concept deeply? What assessments do you rely on and what do they tell you? What do you notice, informally, and how do you interpret your students' actions regarding their creation of meaning? What might you see (or not see) that suggests your students have a rich understanding of the math at hand?*
- *What do you do to foster meaning when you feel that time is limited or that there is too much to do in each math lesson? How do you balance the process of helping students develop meaning with the pacing of your curriculum or math program?*
- *What do you do in your classroom to promote perseverance? What do you do when students give up on a task? How do you help your students develop mathematical resilience and a willingness to stick with an activity?*
- *How well do your activities and lessons promote and foster perseverance? To what extent do they draw students in and engage them in mathematical thinking in a way that is propulsive?*

References

Aristotle. (1925). *Aristotle in* 23 Volumes (H. Tredennick, translator). Cambridge, MA: Harvard University Press.

Brownell, W. A. (1947). The place of meaning in the teaching of arithmetic. *Elementary School Journal, 47*(5), 256–265.

Carter, S. (2008). Disequilibrium and questioning in the primary classroom: Establishing routines that help students learn. *Teaching Children Mathematics, 15*(3), 134–137.

Clements, D., & Sarama, J. (2009). *Learning and teaching early math: The learning trajectories approach*. New York: Routledge.

Jacobs, V. R., Lamb, L. L., & Philipp, R. A. (2010). Professional noticing of children's mathematical thinking. *Journal for Research in Mathematics Education, 41*(2), 169–202.

King, N. L. (2014). Perseverance as an intellectual virtue. *Synthese, 119*(15), 3501–3523.

Lannin, J., Chval, K., Dougherty, B. J., & Jones, D. (2013). *Putting essential understanding of multiplication and division into practice, in grades 3–5*. Reston, VA: National Council of Teachers of Mathematics.

MacDonald, B. L. & Thomas, J. (2023). *Teaching mathematics conceptually: Guiding principles for 5–10 year-olds*. London: Corwin UK.

Mayer, R. E., & Hegarty, M. (1996). The process of understanding problems. In R. J. Sternberg & T. Ben-Zeev (Eds.), *The nature of mathematical thinking* (pp. 23–53). New York: Routledge.

National Governors Association (NGA) Center for Best Practices & Council of Chief State School Officers (CCSSO). (2010). *Common core state standards for mathematics*. Washington, DC: Authors.

Reinhart, S. C. (2000). Never say anything a kid can say. *Mathematics Teaching in the Middle School, 5*(8), 478–483.

Steffe, L. (1992). Learning stages in the construction of the number sequence. In J. Bideaud, C. Meljac, & J. Fischer (Eds.), *Pathways to number: Children's developing numerical abilities* (pp. 83–88). Mahwah, NJ: Lawrence Erlbaum.

Su, F. (2020). *Mathematics for human flourishing*. New Haven, CT: Yale University Press.

Thomas, J., & Tabor, P. D. (2012). Developing quantitative mental imagery. *Teaching Children Mathematics, 19*(3), 174–183.

Thomas, J. (2017). The ascendance of noticing: Connections, challenges, and questions. In E. O. Schack, M. H. Fisher, & J. Wilhelm (Eds.), *Research trends in mathematics teacher education* (pp. 507–514). New York: Springer.

Vygotsky, L. S. (1978). *Mind in society: The development of higher psychological processes*. Cambridge, MA: Harvard University Press.

Webb, D., Boswinkel, J. G. H., & Dekker, G. H. (2008). Beneath the tip of the iceberg: Using representations to support student understanding. *Mathematics Teaching in the Middle School, 14*(2), 110–113.

Wright, R. J., Martland, J., & Stafford, A. (2006). *Early numeracy: Assessment for teaching and intervention* (2nd ed.). London: Paul Chapman Publications/Sage.

3

Play and Exploration

For each chapter, I have curated a playlist from my own music library. These are songs that put me in the mindset of the topics at hand. Some may be on the nose and others may feel like a bit of a stretch. Hopefully, though, you find the music an enjoyable addition to your thinking in these areas.

Note: Spotify automatically adds selections to a given playlist. My original selections are the first 6–8 songs on a given list

https://open.spotify.com/playlist/0DVysOhUwo6IFou6kz52vT?si=BWs48bZ4RIKdnkQ6It-syQ

Ethan is Sick of Math

Most of the stories I share in this book are from my own classrooms or those of my teaching friends. However, to begin our journey into play and exploration, I thought a personal anecdote involving one of my own kiddos, Ethan, could be illustrative. At the time, Ethan was in second grade and, by all measures, an absolute wiz at math. Whether it was state testing, quizzes, or classroom progress monitoring, Ethan was crushing it in math class. As a longtime skeptic of standardized tests, I have to admit that I took pride in Ethan's math performance. Though I was dubious of the value of rote fact memorization, I remember

DOI: 10.4324/9781003484783-3

FIGURE 3.1 Ethan's Math Fact Crown

smiling when he would bring home another math fact crown (see Figure 3.1).

Given my own struggles with math as a young person and my later love of all things mathematical, I was pleased to see my son seemingly taking to math so readily. You can imagine my surprise, then, when I picked him up from school and he told me how he absolutely couldn't stand math. "I am just sick of it, Dad! I don't want to do it. Do I have to keep doing it?". He went on to explain that his math class that day had been built around yet another series of worksheets (ostensibly aimed at developing procedural fluency) and he was just so, so bored with doing the same types of tasks over and over.

At this point, I want to take a moment to share my perceptions of Ethan's teacher, let's call her Ms. Roberts. Throughout the year, by Ethan's own account, Ms. Roberts had been a phenomenal teacher. Most days, he was genuinely excited for school, and the stories he told about his lessons suggested a teacher who was very invested in creating inclusive and interesting spaces for meaningful learning. My partner and I had similar thoughts about Ms. Roberts. When we met for parent/teacher conferences, she knew Ethan deeply as a young person. She seemed to really "get" our kiddo, and our sense was that she had that sort of relationship with all of her students. In short, she struck us as a highly invested and capable teacher, and we thought very highly of her. However, when it came to mathematics teaching and learning, she prioritized procedures and fact recall, which resulted in classroom experiences that were, at least for Ethan, far less engaging than other portions of the day. There are several different and important reasons this might have been the case, which we will take up later in this book. Germane to this anecdote, I want to make clear that we were all fans of Ms. Roberts, Ethan included.

With that backdrop in mind, it was all the more frustrating to learn that Ethan was becoming disillusioned with math. On that ride home, I asked him what he wished would happen in math class. I shared with him the need to learn how numbers work, to be able to look for patterns, and to use all of these skills to solve important problems. I shared with him that not everything in school is going to be enjoyable and sometimes we just have to do the work that's in front of us. He seemed to take all of this in, and then posed a question that, in large part, led me to write this book. *"Why can't we do both?"* Ethan, through his second-grade lens, was wondering why teachers couldn't foster mathematical learning while simultaneously providing students with an engaging experience. Why can't we do both, indeed!

Arriving home, I knew that day was my turn to make dinner, but I didn't want to lose this moment. Right then and there, I wanted Ethan to have a mathematical experience organized around a key concept that provided opportunity for meaning and perseverance (see Chapter 2), and perhaps most importantly, that was fun. On a shelf in the basement, I saw the game, Katamino (1992), pulled it down, and began sharing the rules with Ethan. If you are not familiar with Katamino (see Figure 3.2), imagine the video game *Tetris* made of wood, and you would have the general idea. This type of polygonal tiling game, often referred to as Pentaminoes, actually has a long history and has puzzled minds for over a century (Dudeney, 1908).

While there are many different variations of Katamino gameplay, one of my favorites is dividing up the wooden pieces

FIGURE 3.2 Katamino (1992)

equally and then taking turns placing a piece on the game board. The goal is to place as many of your own pieces as possible and to prevent your opponent from placing pieces. Since players can see each other's blocks, both can think strategically about how best to place their respective pieces. The person who places the last piece wins the game. This game draws upon spatial reasoning and how best to position one's blocks in concert to allow for the placement of as many of one's own blocks while simultaneously foreclosing on space for opponents to place her/his/their blocks. This sort of mathematical play allows for tangible exploration of key ideas such as rotation and reflection that are formalized in middle and secondary settings. For example, the authors of the Common Core State Standards for Mathematics, in the section describing key high-school concepts, write,

> [t]he concepts of congruence, similarity, and symmetry can be understood from the perspective of geometric transformation. Fundamental are the rigid motions: translations, rotations, reflections, and combinations of these, all of which are here assumed to preserve distance and angles (and therefore shapes generally). Reflections and rotations each explain a particular type of symmetry, and the symmetries of an object offer insight into its attributes—as when the reflective symmetry of an isosceles triangle assures that its base angles are congruent.
>
> (p. 74)

Playing Katamino provides a real and engaging foundation for these more formalized concepts that students may build upon as their mathematical understanding in this domain develops.

After playing a few rounds, I asked Ethan what he thought of the game, and he said it was pretty fun, but also challenging. I asked him what made it challenging, and he said (paraphrasing from memory),

> It's hard to figure out how to use my blocks so that they fit together but also take up more room so that you can't put your blocks down. You have to kind of imagine them in a bunch of different ways at the same time and then imagine all the different ways you could play your blocks. Then you just have to make a good guess of what your best block is and how to set it down. It's a lot to keep in your head.

As I heard him talk about what made this game difficult, I thought that this was precisely what made it engaging and fun as well. His imagining of all the different ways his blocks might rotate or reflect on the board as well as the ways I might rotate and reflect one of my blocks in response, was engaging in a way that the math of his school worksheets wasn't. In this moment, we were using our mathematical minds and playing against one another, but there are other variations where we might have played collaboratively. In any event, exploration and play were central to the experience.

Now, at the end of this lengthy anecdote, if you are saying to yourself, "that's a wonderful story, but as a teacher, I can't just pull a game off a shelf and hope my students learn a key math concept", I understand and relate deeply. Mathematics is a unique playground and developing an understanding of how all the concepts, processes, and structures work requires some specialized experiences that can't always be ported to a puzzle or game. Rather, what we aim to explore in this chapter is how to overlay a *spirit* of play and exploration to our teaching and learning such that students are drawn into the wonder of mathematics rather than repelled by it.

Let's Play

My Mawmaw is a funny lady. When I ask her a question, she usually answers in the craziest way possible. One day, when I was getting ready to visit her, I asked her who all was at her house, and she said, "Well, there are *12 legs and 2 tails* in my house not including me". What could be in Mawmaw's house?

Jot down your ideas here. Share them with your students. See if they can build on your ideas.

____

____

____

____

The Virtue of Play

In the first chapter, I introduced play as a key virtue that might shape the experiences in our mathematics classrooms. As a broad theme, I asserted that play is an expression of our human selves and that such play can and should be very fun. Let's go a step further and consider play through historical and mathematical lenses. Looking back to antiquity, as one might imagine, the human pursuit of play fascinated thinkers and philosophers. Aristotle (trans. 2002) explored the concept of playfulness (e.g., wittiness, absurdity, and lightheartedness), termed *eutrapelia*, and sought to explore how play might be positioned and braided with other virtues to pursue a full existence. Despite a sustained fascination with the experience of play, scholars often struggle to define exactly what play is (Eberle, 2014). In his seminal work, *The Ambiguity of Play*, Sutton-Smith (2001) writes,

> We all play occasionally, and we all know what playing feels like. But when it comes to making theoretical statements about what play is, we fall into silliness. There

> is little agreement among us, and much ambiguity . . . Communication theorists tell us that play is a form of metacommunication far preceding language because it is also found in animals. Sociologists say that play is an imperial social system that is typically manipulated by those with power for their own benefit . . . Anthropologists pursue the relationships between ritual and play as these are found in customs and festivals . . . In psychiatry, play offers a way to diagnose and provide therapy for the inner conflicts of young and old patients alike. And in the leisure sciences, play is about the qualities of personal experience, such as intrinsic motivation, fun, relaxation, escape, and so on.
>
> (pp. 1–7)

Despite this deep ambiguity, often driven by one's perspective and positioning with respect to play, Sutton-Smith presents several core *rhetorics* that help us describe fundamental aspects of play. For example, the rhetoric of play as progress describes the way that playful activity can lead to social and cognitive growth. Similarly, the rhetoric of play as the imaginary "idealizes the imagination, flexibility, and creativity of the animal and human play worlds" (p. 11). Each of these rhetorics (i.e., self, frivolity, identity, power) may be used as a lens through which we can better understand play as a core human endeavor.

Alison Gopnik, renowned child psychologist and the first author of *The Scientist in the Crib* (Gopnik, Meltzoff, & Kuhl, 1999), has conducted considerable study of play among infants and young children. Throughout myriad experiments, she found that babies and children have a profound desire to play. From social games involving groups of young children, to individual infants manipulating objects to produce exciting and novel outcomes, play is very important. Likely, anyone who has spent time around children will agree that the seeking of play and engagement in play (whether it be at a park or on an iPad) is a big deal for these young people (and most adults, too). Regarding that rhetoric of play and progress, though, Gopnik (2009) offers some cautionary advice,

> Sadly, some parents are likely to take the wrong lessons from these experiments and conclude that they need programs and products that will make their babies even smarter. Many think that babies, like adults, should learn in a focused, planned way. So parents put their young children in academic-enrichment classes or use flashcards to get them to recognize the alphabet. Government programs like No Child Left Behind urge preschools to be more like schools, with instruction in specific skills.

Rather than using our increasing knowledge of human development to create more rigidly structured pathways of progress, Gopnik argues for opening educational systems to accommodate more play as the vehicle for cognitive and physical development, for invention, for enhancing their human experience, and to have lots and lots of fun.

Given our understanding that play, by its very nature, has a cognitive axis, there is considerable space to cojoin such play and the world of mathematics. On this point, Su (2020) writes,

> Mathematics makes the mind its playground. Doing math properly is engaging in a kind of play: having fun with ideas that emerge when you explore patterns, and cultivating wonder about how things work. Math is not about memorizing procedures or formulas, or at least that's not where you start . . . Math play is voluntary but driven by a deep curiosity that is nurtured through practice – much like how the more you get a feel for a new game, the more you want to play. The most practiced adherents of math play will see a new idea and feel a compulsion to play with it.
>
> (pp. 51–53)

Su continues to describe mathematical play as having two phases: *pattern exploration* and *inductive reasoning*. The first phase of pattern exploration involves considering some mathematical idea and what we find noteworthy or interesting. Indeed, math teachers attuned to drawing on such playful posture often ask students, "What do you notice? What do you wonder?" Returning to the math moment in Chapter 2 that centered on the pattern $5^2 = 25$, $15^2 = 225$, $25^2 = 625$, and so forth (see Figure 2.3), the teacher begins that pattern exploration with the prompt, "Think about this for a few minutes. What jumps out at you? What are some things you notice? What do you start to wonder? What connections can you make?" The key to this phase is to open up space for students to play with ideas in their minds – to take their thinking in new and novel directions. Sometimes this play might travel to unusual places. For example, imagine a student plays with the size of the exponent notation (i.e., enlarging the superscript) and contemplates $52 = 25$ (rather than $5^2 = 25$). On the face, this is a nonsensical expression – 52 most certainly does not equal 25; however, if we take a step back and consider the relationship between these two numbers (52, 25) – the reverse of one another, then a door opens up to a whole new world of patterns and sequences. What operators and numbers might we place between 52 and 25 to make a true mathematical statement? Using what we know about the rules of the game to see where our mind can take us leads us toward mathematical play.

Delving deeper into such play, we begin inductive reasoning. In this phase of play, the questions driving our playfulness move from what we notice and wonder to *why* these things we notice and wonder are so. Su (2020) writes, "to show the truth of all statements in a sequence, an experienced math explorer might try a *proof of induction*, which uses the truth of one statement to deliver the truth of the next, the way a line of dominos fall when you tip the first one. Such a collection of opening moves for a math proof is just like a catalog of opening moves for a game of chess. They'll get you off to a good start" (pp. 55–56). What Su is describing, here, is the way that play moves from the exploratory to the strategic. My kiddos like it when I play video games with them because most of the time, I am simply mashing buttons to see what my character can/will do; however, they have a much more strategic sense of how to use controls to achieve a goal, which in the case of Nintendo's *Mario Kart*, is to completely humiliate me on the racetrack. However, I remember their first experiences with games, and that play was a lot more exploratory in nature – What does this button do? What if I press these two buttons together? What if I jump onto this flashing platform? As they gained experience and understanding, their play moved from exploration to more strategic reasoning, and the best designers know how to create engaging pathways for this transition. This is the sort of experience I am proposing for our math classrooms. Indeed, if we use the old saw of beginning with the end in mind, let's consider how might we take a mathematical goal (e.g. standard, and learning target) and play our way toward it in both a fun and increasingly strategic manner. On this point, it is worth mentioning that most research describes it as a voluntary process, and that we engage in play on our own volition and preference. On the surface, this might seem at odds with the compulsory nature of math class and schooling more broadly. However, if we take this up as a design challenge in the same manner as creators of other play media and tools (e.g., game designers and toy designers), our work as teachers moves from "how might students best learn this concept" toward "how might students play with this concept in a most engaging way". We leave the role of *math-explainer* behind and take on the role of *experience-designer*, and therein lies all the difference as our work becomes our play.

The Virtue of Exploration

If you noticed the word "exploration" pop up frequently in the previous section, that was no accident. As we know, it's difficult to consider the virtue of play without touching on its exploratory nature. The playful questions of "What do you notice?" and "What do you wonder?" are also deeply exploratory. Su (2020) referred to his first phase of mathematical play as pattern exploration. Hence, we will examine the virtues of play and exploration in concert.

Often when we think of exploration, one imagines the seeking travelers of the past – Marco Polo traveling the silk road, or perhaps Sacagawea aiding Lewis and Clark's expedition across the lands of the Lakota and Sioux nations. However, exploration need not be considered through some distant or lofty lens. One can just as easily explore a park in a neighborhood, an idea in a book, or, via reflection, an event from one's past. No matter the context or occasion, a central feature of exploration is curiosity. The origin for all exploratory activity (assuming it is voluntary), is a deep curiosity of the unknown, and we can turn our exploratory interests inward as well as outward. What things might exist behind the rock outcropping in the park? What was Fitzgerald trying to tell us with that green light at the end of Daisy Buchanon's dock in *The Great Gatsby?* What really happened in my early college courses that made me fall in love with mathematics when I had disdained it for so long? For each of these questions, one seeks to see or better understand something currently unknown. Moreover, as a core driver of exploration aimed at some human fulfillment, curiosity, in its positive manifestations (rather than, say, morbid curiosities), has deep virtuous connections. In his essay on curiosity as a moral virtue, Baumgarten (2001) writes,

> It seems fair to conclude that curiosity is a distinctive virtue which, compared to attentiveness and "being interested," more fully expresses human autonomy, plays a distinctive role in caring relationships, and enables us to learn about things we would not otherwise know. Insofar as curiosity contributes to caring, it is likely to have social benefits, but since caring and close relationships are reasonably considered components of a well-lived life, curiosity has moral value for a person even apart from its contribution to the welfare of others.
>
> (p. 175)

What Baumgarten skillfully argues throughout this essay is that, when aimed in positive directions, curiosity and our subsequent desire to act on such curiosity via exploration carries inherent virtue and may move us toward unique fulfillment as human beings. From infancy on, our curiosity has the power to push us to discovery and growth. "[B]abies and young children are exquisitely designed by evolution to change and create, to learn and explore" and this facet of our human nature remains with us throughout our lives so long as we choose to engage with it (Gopnik, 2010, p. 81).

As you might suspect, being a willing explorer in mathematical realms can also be remarkably fulfilling. Su (2020) writes,

> Mathematical exploration is very much like space exploration, but of a different kind of space – a space of ideas. You don't know what you will find when you start. You send out probes to test theories. You are captivated by mystery, motivated by questions, undeterred by setbacks. You make discoveries from a distance: because the ideas themselves are not physical, you access this space through reason. Exploration and understanding are at the heart of what it means to do mathematics.
>
> (pp. 22–23)

Exploring mathematical ideas can be richly satisfying. Young children may be inherently fascinated by the vastness of quantity and wonder how far they might count in search of the *biggest number*. Older individuals might marvel at the equality. 99999999 [repeating] = 1 and how that might be so. Intuitively, it feels as though that decimal (stretching all the way to infinity) would never quite reach 1, but it does. That curiosity of why this is so could drive some rewarding investigation of various proofs. One simple entry point might be to consider how the fraction 1/3 is expressed as a decimal and what happens when we multiply this decimal expression by 3 and comparing it with multiplying 1/3 × 3 in fractional form (Byers, 2007). There are nearly limitless mathematical places and spaces where some curiosity might result in exploration. Perhaps that exploration comes with satisfying resolution or perhaps not, but merely engaging with and acting upon one's curiosity to investigate a mathematical unknown can be uniquely virtuous and rewarding to us as human beings. In the next sections, we will examine, precisely how we might engage one another in play and exploration to engender a rehumanized mathematical experience.

Let's Explore

Consider the problem 5 ÷ ¼. Try to create a word problem that fits with this task. What does it mean to divide by a fraction? Where might that process exist in the world around us?

Jot down your ideas here. Share them with your students. See if they can build on your ideas.

____

____

____

____

Building a Culture of Play and Exploration

As teachers, there are many ways we might foster the virtues of play and exploration in our classrooms. Arguably, though, the choices we make about the mathematical tasks and activities that ground our instruction set the stage for how our students will experience a concept or idea. If we choose more routine problems, our students will likely have a more routine experience. If, however, we identify, create, or adapt existing resources in more novel and open-ended directions, then students may have a more novel and open-ended experience. Recall from the preceding chapter, Mayer and Hegarty (1996) write,

> A routine problem exists when a problem solver knows how to carry out the correct solution procedure and recognizes that the solution procedure is appropriate for the problem . . . Routine problems are not really problems at all because the problem solver knows what to do and how to do it. For this reason, such problems are often referred to as *exercises*. A non-routine problem exists when a problem solver has a problem but does not immediately see how to solve it.
>
> (p. 32)

What these authors are describing is the way that the mathematics classroom will often foreclose on opportunities for play and exploration in favor of routine exercises. The prevailing notion in these instances is that this is simply the way that mathematics must be taught and learned. However, we know from research that true problematic tasks are more fruitful for conceptual growth; moreover, we also know from research that play is essential to the human experience and a driver of persistence within an activity. With these ideas in mind, what shape might we give to tasks and activities in our classrooms to foster play and exploration? Below are some examples that I hope provide an entry point for you to think about building your own classroom culture around these virtues.

Imagine our goal was to work with students on verbal counting sequences. Such verbal sequences are an essential element of early counting activity (Thomas, Tabor, & Wright, 2010) as well as being supportive for arithmetic operations including multiplicative reasoning via skip-counting sequences. Note, the Common Core State Standards (2010) also make explicit reference to verbal sequences as key learning goals. There are many ways we might lead students in activities around verbal sequences. For example, we might ask our students to get in a circle each individual says a number as we verbally count forward (or backward) around the circle. Or, we might think about ways to approach this goal more playfully. Music can sometimes be a good portal into a more playful space. Consider the nursery rhyme, *Twinkle, Twinkle Little Star* (Figure 3.3).

Now, if you are like me, then you might be stuck hearing a snippet of this melody in your head for the remainder of the day. For some reason, this little song has always been quick to stick in my mind. However, these sticky little musical structures, or *earworms* (Beaman, 2018), can be a playful way into verbal sequencing work. Perhaps we task students with *singing* their sequence to a popular song using that song's melody and tempo (see Figure 3.4).

In this instance, notice how the sequence becomes more difficult as we reach some of the (multisyllabic) teen numbers (11, 13, 14). The syllables of the counting sequence strain the

Tween – Kul – Tween – Kul
Lit – Tul – Star
How – I – Wun – Dur
Wut – You – Are

FIGURE 3.3 Twinkle Twinkle Little Star (Phonetic)

1 – 2 – 3 – 4
5 – 6 – 7
8 – 9 – 10 – 11
12 – 13 – 14

FIGURE 3.4 Verbal Sequence

tempo of the original song. Arguably, though, that's another element of play as we struggle to match our counting words with the song. Popular music is also a great source for merging this sort of counting work with song. For example, the introductory chant to Lady Gaga's *Bad Romance* (rah, rah . . . ah, ah, ah!) is a particularly sticky melodic rhythm and great for counting along. If we want to layer in exploration, perhaps a task element is for students to try and find a song that best fits, rhythmically, with a given verbal sequence. In any event, we have moved what might be a more mundane counting experience into a more playful arena.

Games are another way to bring play and exploration into our classrooms. As I mentioned earlier in this chapter, I looked to the game *Katamino* (1992) to reorient my son into a world where math and play may be cojoined. There are plenty of other well-designed games that have a strong mathematical connection that we might use to bring our teaching beyond the routine. Stroll through the game aisle of Target or Wal-Mart and you will find many different games that connect, in some manner, to an important mathematical concept. *Quirkle* (MindWare, 2006) and *Rummikub* (Pressman, 1998) are very engaging games predicated on pattern building and pattern recognition. Any game that requires moving pieces along a board according to a dice roll is supportive for early counting activity. More sophisticated games such as *Ticket to Ride* (2004) and *Catan* (1995) require coordinating quantitative reasoning with prediction to manage and strategically deploy resources. With a little thought and planning, we as teachers can position a rich and well-designed gaming experience at the center of our math lessons and use that experience to help students construct important concepts and ideas through play and exploration.

Lastly, to build such a culture, we as teachers may act as models of these virtues. For reasons unknown to me, our society seems to have created some strong associations with math and seriousness. It needn't be this way. As teachers of mathematics, we can embody and model a spirit of playfulness for our students. As a new teacher, I was given a lot of coaching to establish authority in my classroom – to maintain control via various management strategies. Decades later, I wouldn't use the word authority to describe my teaching approach, but I do understand the importance of effectively managing an instructional experience. However, this doesn't mean that we can't approach our instruction with a sense of play and whimsy. If we are tasking students to create a counting sequence song, let us teachers sing along too. If a moment calls for movement or dance, let us teachers move and dance too. If we are leading a verbal counting sequence, grab a pencil, use it as a conductor's baton, and turn it into a counting symphony! Find opportunities to infect your lessons with smiles and laughs, and your students will mirror those emotions. In times of exploration, model uncertainty. Even when we have a clear vision of the math at hand, we might dial our language back a bit to reflect a more tentative posture or travel pathways we know to be unfruitful. Returning to an example from earlier in this chapter, the Mawmaw problem, we might say, "Well, she might have a spider, so that's 8 legs and no tails, what could I put with that? A dog? No, that doesn't work, because it's the right number of legs, but not enough tails". What we are doing in these moments is lowering the stakes of exploration. So often, our students look to us as the *keepers of right and wrong*, and even when they are confident in their own thinking, they seek external validation of their thinking from us. In a classroom organized around play and exploration, we want them to see a leader who is willing to walk that path too. Someone who, while in charge of the moment, is willing to peek around blind corners and walk toward places unknown and have a great deal of fun doing so.

Core Practice: Playing the Believing Game

While there are many ways that we, as teachers, can support our students' engagement in play and exploration, arguably a foundational practice is in our ability to *play the believing game* (Elbow, 2006; Harkness, 2008). In a nutshell, this practice is aimed at leaning into our students thinking, trying to find some truth in their thinking, and stepping back from our own notions of mathematics a bit to *believe* in what our students are sharing with us. Harkness writes,

> In mathematics we tend to value the talent to doubt, to criticize, and to find flaws with students' thinking or understanding. We use one counterexample to prove that a conjecture cannot be true. We try to find the errors and misconceptions in our students' computation or reasoning . . . In order to try to believe, mathematics teachers must suspend their own logic, at least until they attempt to first understand and honor the logic of the student.
>
> (p. 245)

What Harkness describes is not an abandoning of our (presumably sound) mathematical ideas and concepts, but rather a generous curiosity to consider the ideas of our students, no matter how outlandish on the surface, and search for some truth therein.

Years ago, I was having a conversation with a student about a meter stick. I called attention to the numbers on the meter stick and asked, "where are the small numbers on the stick"? The student touched an area near the 2 cm mark. I then asked, "So, where are the large numbers?" The student looked at me quizzically for a moment, and said, "All of the numbers are small". It was immediately apparent to me that this student and

I were using the words *small* and *large* differently. While I had hoped the student would make some comparison quantitatively about the magnitude of the numbers at one end of the stick vs. the other, he was thinking more about the physical size of the numerals themselves. More importantly, stepping away from my magnitude-oriented goal, we can easily see the truth in his position. Fundamentally, this is what playing the believing game is all about – trying to see the mathematical situation through our students' eyes and see their truths therein.

These opportunities to play the believing game abound in our mathematics classrooms if only we train our eyes and minds to look for them. Harkness et al. (2009) recount a novice teacher's experience working in mean, median, and mode. This teacher, named Amber, when looking back on this moment, discovered an important opening to believe in her students' mathematics:

> I asked my students to create books about mean, median, and mode. Students were asked to define each type of average, highlight important features, and provide an example from everyday life. In Debbie's [student] description of mode she explained that mode is 'the number used the most', and she used this strategy to *determine the answer of 4* [emphasis added] in her example. Debbie drew pictures of things she would see at a park and put the number of these items at the park under each picture . . . I am sure that at the time I graded this project I thought, 'This is incorrect'. The mode of this situation is actually the trees, because there were 8 of them at the park [see Figure 3.5]. That is the item that occurs most frequently. Actually, my thinking was limited and absolute. I needed to remember that numbers and items reside within the definition of mode depending on how the situation is interpreted.
>
> (pp. 40–41)

In this retelling of events, Amber describes the tension between her perspective of mode and that of her student's. From Amber's perspective, *trees* is the correct answer as it represents the

FIGURE 3.5 Debbie's Drawing

numerical mode – the most frequently occurring *type* of item in the park. However, from the student's perspective, *four* is the correct answer as it represents the most frequently occurring *number* of items in the park. Practicing the believing game allows us to sit with this student's ideas, consider if there may be some mathematical truth within, and then work toward some shared understanding between student and teacher. Within the field of improvisational comedy, there is a guiding principle for performers to never say no. That is, to build upon or lean into other performers words and actions to keep the comedy moment viable. Never saying no allows comedians the space to expand ideas, play with and explore them in real time, and almost always generate a far funnier scene. In essence, this is the believing game as applied to comedic performance. Training ourselves to play the believing game with students allows us to keep a mathematical moment alive and to maintain a playing field where students feel comfortable letting their guard down, engaging in play, and exploring ideas heretofore unknown.

Professional Noticing for Play and Exploration

As with most every aspect of mathematics teaching and learning, the practice of professional noticing can profoundly shape the experience of our students and that is certainly true in the domains of mathematical play and exploration. Indeed, professional noticing serves as a connecting practice across most all of our teaching endeavors. Recall that professional noticing is comprised of three elements: attending, interpreting, and deciding (Jacobs, Lamb, & Philipp, 2010). In core, we observe a mathematical moment closely, gathering in as much detail as possible (attending), then we draw some meaning from what we heard and saw (interpreting), and based on how we understand that moment, arrive at some response or action (deciding) – see Chapter 2 for more detail on this practice. With those components in mind, consider the following small-group exchange around place value (see Figure 3.6).

Attending to the words and thinking in this exchange, we first see the teacher moving the task into what might be considered unconventional place value space (i.e. 300 hundreds). Subsequently, there is some initial rejection of the premise by Taylor ("That's not right".) and continued grappling of this idea by Willem ("But you still can't write three hundred hundreds"). The teacher endeavors to keep space open for this thinking as well as leaning into mathematical novelty ("Maybe it sounds a little weird, but in math, we can write the same number in all sorts of crazy ways"). From there, Josiah appears to grasp the

TEACHER: *"So thinking about our place value, if I give you a description of a number, do you think you could figure out what that number is? Warning, I might do this one a little crazy"* [writes on the board] "300 hundreds; 35 tens; 50 ones"

TAYLOR: *"Wait, that's not right. Can you do that?"*

TEACHER: *"Maybe. What do you think? Is there a number in there?"*

JOSIAH: *"Sort of, yeah, I think so. Like, thirty-five tens is three hundred fifty, so we know that piece"*

TAYLOR: *"But three hundred hundreds doesn't make sense. It's like. . ."*

TEACHER: *"What's ten hundreds?"*

WILLEM: *"That's a thousand. But, one hundred hundreds, that's not . ."*

TAYLOR: *"No, a hundred hundreds is ten thousand, right? It's just taking it to the next level. Like a thousand times ten again."*

JOSIAH: *"Right, so three hundred hundreds is thirty thousand, so now it's thirty thousand, three hundred fifty and then the fifty ones."*

WILLEM: *"But you still can't write three hundred hundreds. It sounds weird and doesn't make sense."*

TEACHER: *"Maybe it sounds a little weird, but in math, we can right the same number in all sorts of different ways. Place value is about how our number system is all about tens, but we could write all sorts of crazy stuff and still figure out a value."*

JOSIAH: *"Yeah, it's not the normal way to write it, but you could write it that way. Like if I wanted to say something like twenty fifteens plus nine sevens. It's weird, but I could figure out that ten fifteens is one hundred fifty, then double it to three hundred."*

WILLEM: *"You just broke my brain with twenty fifteens."*

TAYLOR: *"No it works, it's just weird. I could say a thousand sevens, and it would be seven thousand. But the normal way is to say seven thousands, right? But I can do that with anything. Three million nines. Seventy-four twelves."*

TEACHER: *"Quick, what's the weirdest way you could write the number five thousand, three hundred, twenty-five using whacko place values? I'll play too. Take two minutes."*

FIGURE 3.6 Unconventional Place Value

possibility of the ideas in play ("Like if I wanted to say something like twenty fifteens . . .") and Taylor seems to have something of a breakthrough ("No it works, it's just weird"). Moving into interpretive space, we might consider how canonical place value (ones, tens, hundreds, etc.) has been constructed by some of these students in a fairly rigid manner (i.e., Taylor and Willem) and how such rigidity might work against a more fluid quantitative flexibility. Indeed, 300 hundreds is unconventional, but it does represent a value – and one that is determinable assuming an understanding of conceptual place value. Lastly, in terms of deciding, the teacher appears to call something of an instructional audible with an improvised task ("Quick, what's the weirdest way you could write the number . . ."). In this moment, the teacher, as a member of the mathematical community sensed a willingness among students to engage with this novelty and leaned into it with a new task. From an improvisational comedy perspective, the teacher said "yes" to the moment and just went with it. This decision, ostensibly, opens up more room for play and exploration among students as they train their thinking in unconventional directions – even if for a few minutes. For us, as teachers, engaging in professional noticing in our mathematical moments is essential to capitalize on such opportunities. Attending to the words and actions of our student allows us to capture detail. Looking into that detail so we better understand their thoughts and emotions gives us a sense of whether they are willing to play and explore. And then, we might formulate some decision that opens a door into a new, engaging, and playful space. This sort of responsive teaching is at the heart of creating a mathematics classroom build around play and exploration.

Questions for Reflection

At the end of each chapter, I will present some questions that, hopefully, drive some reflection, introspection, and perhaps conversation among you and your colleagues. Given the richness and complexity of the ideas we have explored in this chapter, it is useful to create some contemplative space for our own thinking.

These questions do not have easy answers, but my hope is that they propel you in a productive direction and keep you thinking hard about play and exploration in the math classroom.

- *How often do you engage in mathematical play or explore new mathematical ideas? If it's not that often, is this something you would like to do more of? Is this something that you shy away from and why?*
- *When you plan a lesson or math experience, do you think about students' opportunities to play with ideas, approach them in a novel way, or explore by taking a concept or process in their own direction?*
- *Do you feel empowered or that you have the freedom as a mathematics teacher to open spaces for play and exploration? If so, where in your curriculum materials do you find those openings? If not, what structures keep you from doing so?*
- *Do you feel that your school culture is one that values play and exploration? Are these themes that students will encounter in other school spaces? Do you have colleagues that think similarly about play and exploration in their own teaching? How does your principal or school leader feel about play and exploration in teaching, and what that means for mathematics teaching and learning?*
- *As human beings, we are not always in a playful mood and exploring unknown things can often be scary. What should we do if a student doesn't wish to play or explore? How might be provide opportunity without forcing compliance with respect to play and exploration?*

References

Aristotle. (2002). *Nicomachean ethics* (S. Broadie and C. Rowe, translators and commentary). New York: Oxford.

Baumgarten, E. (2001). Curiosity as a moral virtue. *International Journal of Applied Philosophy*, *15*(2), 169–184.

Beaman, C. P. (2018). The literary and recent scientific history of the earworm: A review and theoretical framework. Auditory Perception & Cognition, 1(1–2), 42–65.

Byers, W. (2007). *How Mathematicians Think: Using Ambiguity, Contradiction, and Paradox to Create Mathematics*. Princeton, NJ: Princeton University Press.

Catan [Board Game]. (1995). Catan Studios. https://www.catan.com

Dudeney, H. E. (1908). *The Canterbury puzzles and other curious problems*. New York: E.P. Dutton.

Eberle, S. G. (2014). The elements of play: Toward a philosophy and a definition of play. American Journal of Play, 6(2), 214–233.

Elbow, P. (2006). The believing game and how to make conflicting opinions more fruitful. In C. Weber (Ed.), *Nurturing the peacemakers in our students: A guide to teaching peace, empathy, and understanding* (pp. 13–25). Portsmouth: Heinemann.

Gopnik, A., Meltzoff, A. N., & Kuhl, P. K. (1999). *The scientist in the crib: What early learning tells us about the mind*. New York: HarperCollins.

Gopnik, A. (2009, August 15). Your baby is smarter than you think. *New York Times*. https://www.nytimes.com/2009/08/16/opinion/16gopnik.html?pagewanted=1&_r=2&th&emc=th

Gopnik, A. (2010). How babies think. *Scientific American*, *303*(1), 76–81.

Harkness, S. S. (2008). Social constructivism and the *Believing Game*: a mathematics teacher's practice and its implications. *Educational Studies in Mathematics*, *70*(3), 243–258.

Harkness, S. S., Lane, C. P., Mau, S., & Brass, A. (2009). The believing game in mathematics: Stories in a discipline of doubt. *The Journal of the Assembly for Expanded Perspectives on Learning*, *15*(7), 37–49.

Jacobs, V. R., Lamb, L. L., & Philipp, R. A. (2010). Professional noticing of children's mathematical thinking. *Journal for Research in Mathematics Education*, *41*(2), 169–202.

Katamino [Board Game]. (1992). Gigamic. https://www.en.gigamic-adds.com/game/katamino/index

Mayer, R. E., & Hegarty, M. (1996). The process of understanding problems. In R. J. Sternberg & T. Ben-Zeev (Eds.), *The nature of mathematical thinking*. New York: Routledge.

National Governors Association (NGA) Center for Best Practices & Council of Chief State School Officers (CCSSO). (2010). *Common core state standards for mathematics*. Washington, DC: Authors.

Quirkle [Board Game]. (2006). MindWare. https://www.mindware.orientaltrading.com/qwirkle-a2-32016.fltr

Rummikub [Board Game]. (1998). Pressman. https://rummikub.com and https://www.amazon.com/Pressman-108648-Rummikub-6-Player/dp/B07Q2K1Y33

Su, F. (2020). *Mathematics for human flourishing*. New Haven, CT: Yale University Press.

Sutton-Smith, B. (2001). The ambiguity of play. Boston, MA: Harvard University Press.

Thomas, J., Tabor, P. D., & Wright, R. J. (2010). Three aspects of first-graders' number knowledge: Observations and instructional implications. *Teaching Children Mathematics*, *16*(5), 299–308.

Ticket to Ride [Board Game]. (2004). Days of wonder. https://www.daysofwonder.com/game/ticket-to-ride/

4

Freedom and Power

For each chapter, I have curated a playlist from my own music library. These are songs that put me in the mindset of the topics at hand. Some may be on the nose and others may feel like a bit of a stretch. Hopefully, though, you find the music an enjoyable addition to your thinking in these areas.

Note: Spotify automatically adds selections to a given playlist. My original selections are the first 6–8 songs on a given list

https://open.spotify.com/playlist/0tQgm115v9VYrKv9w78S6N?si=2W5gDMwbSGq-Ko_eeXPviQ

Introduction

When I think back of professional learning experiences that I have had in mathematics education, my recollection is that they were often very structured and directed. Indeed, such focus and intentionality are often considered hallmarks of effective professional development (Wei, Darling-Hammond, & Adamson, 2010); however, many of these well-intentioned trainings did not make me feel particularly independent or empowered. Rather, I was being presented with some new practice, curriculum or technology in a regimented way, and the expectation was that I would develop understanding or mastery of this new thing. I wonder, quite often,

DOI: 10.4324/9781003484783-4

how students feel as they make their way through our intricately designed systems of instruction and assessment. Ostensibly, our schools and classrooms exist primarily to further the intellectual, social, and emotional interests of young people, but when do the students in our care have the opportunity to shape their own experiences? Here's a challenge. Find a fifth grader and ask that individual how often he/she/they feel freedom in math class. Imagine the same experiment with a high-school student. What do you imagine you would hear? How would students describe their freedom, or lack thereof, and the extent to which power is afforded to them? Do you imagine they would have positive things to say about all of this?

Consider the task 32 × 25. Take a moment and work that one out however you would like and describe your strategy in the space below:

Jot down your notes here. Share them with your students. See if they can build on your ideas.

____

____

____

____

____

Perhaps you approached this in writing via the standard vertical algorithm [(2 × 5) + (2 × 20) + (30 × 5) + (30 × 20) = 800]. Maybe you thought, since 3 iterations of 25 equals 75, 30 iterations of 25 equals 750, then I just need to add two more 25s which lands us at 800. Or, possibly, you factored 25 into 5 × 5 and using what you know about 32 × 10, you thought, 32 × 5 = 160, then 160 × 5 = 800. In any event, there are many different, valid ways that we could think about this task. Some of them might appear more efficient than others, but each would contain some mathematical idea that is worthy of our attention and contemplation. Even if I were to add 32 over and over (25 times), there would be opportunities to identify patterns and anticipate interesting outcomes

TEACHER: *"So you all are getting pretty good at multiplication, right? I want you all to try one just in your minds. Try to work out thirty-two times twenty-five just in your head. Any which way you want. I am open to it all."*

CHLOE: *"Can I use my white board?"*

TEACHER: *"Well, except maybe white board. Try to stay just in your head for this one Chloe. I will give you all about five minutes to think this one through and then be ready to talk about your thinking."*

TEACHER: *"OK, looks like most everyone is ready. Who wants to share?"*

LUKE: *"I thought about it like just lining them up. Like, I started with two times five, and then three times five and the rest. I think it's like one hundred sixty plus sixty-four, but that's as far as I got."*

CHLOE: *"I did the same, but it's eight hundred. It's one hundred sixty plus six hundred forty."*

MAYA: *"I started that way, but then I thought about the twenty-fives. That's like a quarter, and four quarters make a dollar. So, I just did that."*

TEACHER: *"What do you mean four quarters make a dollar? How did that help you?"*

MAYA: *"Well, ten twenty-fives is two hundred fifty, then another ten is five hundred. Then another ten is seven hundred fifty. That's thirty twenty-fives. Then you just have two more which is fifty cents, I mean fifty so it's eight hundred."*

TEACHER: *"That's interesting Maya. I want to come back to that in a minute. Did anyone else have a different way of thinking about this?"*

TYLER: *"I kinda thought about it in terms of place value. Like, there are thirty twenties which is six hundred. Then, there's two fives, which is ten. So, I think it's six hundred ten, but I am not sure now."*

TEACHER: *"Ooh, these are really interesting. I would love to hear more, but I have a next step. However you thought about it, I want you to try and draw a picture of what was happening in your mind. Feel free to use models from class like number lines, counters, or that area model we discussed. . . whatever, but try to draw the problem as you were working it out."*

FIGURE 4.1 32×25

along the way – at what number might I expect to arrive after my tenth addition? What about the twentieth addition? How do those two numbers relate to one another? With this in mind, consider the following exchange involving fourth-grade students and their teacher (see Figure 4.1).

In this exchange, we first see the teacher extending freedom of strategy to students (i.e., "any which way you want"); however, there is a key boundary on that freedom in that the terrain for this mathematical activity is to be mental space. This is an important distinction with respect to lending power to students so that they may think and act with freedom in the mathematics classroom. It's easy to dismiss these virtues as a pathway to chaos and ineffective instruction as we imagine lessons being driven entirely by the whims of students who, for the most part, have more limited knowledge of the concepts at hand. Rather, a good goal for the teachers is to begin from a place of maximizing power sharing and freedom and carefully consider what boundaries we might put into place. In this instance, thinking about this task mentally (and distancing from the standard algorithm) was the ostensible goal, so teacher put that guardrail in place. Beyond that, though, the doors were wide open for students to approach this task in any way they chose. Again, the key point here is that power sharing and creating spaces of freedom is not an all or nothing proposition. We, as teachers, reserve the right to layer in guidance and structure, but do so thoughtfully and with an eye toward offering our students autonomy.

Turning to the students' strategies, there was a range of offerings in this moment. Luke and Chloe seemed to mentally work through the standard algorithm with varying results, while Maya thought about the task as iterating 25 in groupings of four (i.e., "four quarters make a dollar"). Tyler, attempted to determine the partial products (i.e., "thirty twenties which is six hundred"), but his process was incomplete. These varying ways of approaching the task represent both the challenges and opportunities of creating mathematical spaces of freedom in our classroom. On one hand, because of the differences and misconceptions, our instructional pathway is not entirely clear. However, the diversity of thought in this exchange also gives us an opportunity to build a broader understanding of 32 × 25. Although they arrived at different answers, Chloe and Tyler could likely have some good conversation about the similarities and differences of their approach – how the standard algorithm produces those partial products. Similarly, Maya's approach (and presumably her subsequent representation) could be a helpful counter to the algorithmic thinking of Luke and Chloe. Similar to our discussion of play in an earlier chapter, the instructional opportunity in this moment is the journey and not the answer. Indeed, it's the varied mental strategies enacted by students that provide us with information about how they are thinking about multiplication as well as a doorway into deepening that understanding via multiple perspectives and paths. Again, lending power to students so that they may freely find their way through a task can land us in a messy space. However, when we critically consider an alternative path where we retain power and freedom is limited, I would argue that there will still be a messy conceptual range of understanding. *We just won't be able to see it*. Consistently responding in unison and following prescribed procedures may produce consistency of behaviors, but it does not produce consistency of thought. Moreover, a classroom that centralizes power with the teacher and limits students' freedom sends a signal that compliance is valued above all else, including one's authentic mathematical thinking.

The Virtue of Freedom

Thus far, we have examined several human virtues (i.e., play, meaning, perseverance, and exploration), and in discussed the interplay among these virtues. For example, play often takes an exploratory form in video games. What happens if I press all the buttons on my controller at once? Holy cow, my character did a crazy backflip kick! Undergirding virtues such as play and exploration, though, is a provision of freedom. Indeed, without the freedom to press all of the controller buttons at once, the extent to which I can explore my video game's features and play in the way that I would like is significantly constrained.

Given its centrality to other virtuous pursuits, concepts of freedom have captivated the minds of thinkers for nearly all recorded history. Biblical accounts of Moses petitioning pharaoh Ramesses for the freedom of the Israelites and Taoist principles of attaining freedom by relinquishing desires and aligning with (rather than fighting against) external constraints (Ni, 1993) are ancient examples of freedom, in its various forms, as something very important to human existence. Although both Plato and Aristotle had somewhat mixed views on the utility of freedom in political systems (Hansen, 2013), they describe *eleutheria* or the opportunity to live as one so chooses, as virtue worthy of pursuit.

Moving forward in time, the Swiss philosopher Jean Jacques Rousseau wrote extensively about freedom and significantly influenced thinkers of his time such as Thomas Jefferson as well as more contemporary reformers such as John Dewey. Rousseau was captivated by the significance of freedom – he once wrote in an informal correspondence to Voltaire, "J'adorè la liberte, Je dèteste ègalement la domination et la servitude" (Eckstein, 1944, p. 264). Loosely translated, this means "I love freedom. I hate domination" and this passion fueled much of his thinking for decades. In his writings, Rousseau explored freedom in many different forms, ranging from civil and democratic freedoms to moral and ethical freedoms (Simpson, 2006). The former pertains to the freedoms expressed by individuals as they form collectives and social contracts and how such organization should always privilege voice and participation of the people involved. The latter refer to the extent to which people experience the freedom to be true to their inner selves and follow their natural callings. In this moral and ethical domain of freedom, Rousseau was careful to draw some boundaries such that freedom could not be used as license for any manner of harmful or destructive behaviors. In a summary of Rousseau's discourses, Eckstein writes, "ethical freedom is by no means identical with lawlessness . . . but is itself based upon a certain kind of law, viz., the law of [a person's] own rational nature" and this rational nature is manifested in some agreed upon rules that allow societies to operate in their most optimal forms (p. 280). At the micro level of the mathematics classroom, this means that we, as teachers, may privilege the freedom of our students so that they may pursue their hopes, dreams, and callings, while still setting some agreed-upon guardrails for the good of the group.

A fun fact about Rousseau is that, in addition to being a brilliant philosopher, he was also an accomplished musician, and he considered music to a "semantic system" for the constructing meaning (Scott, 1997, p. 812). With respect to the ethical and moral freedom necessary in responding to our personal callings, Rousseau thought of this as listening to our own inner songs and being mindful of the music of our souls. Given our focus on virtues throughout this book, it is useful to note the way that great thinkers intertwined powerful elements of the human experience to shape their ideas. How might we create conditions for our students to better hear their inner song and act upon it?

I would be remiss, though, if I left readers with the perspective that philosophies of freedom were the province, primarily, of European scholars. Many hundreds of years before Rousseau was born, Native American tribes such as the Iroquois and Cherokee had well developed notions of freedom and considered, carefully, its virtuous impact on their peoples (Johansen. 1982). For example, the Cherokee invoked directional metaphors (e.g., up and down) and the concept of where one is to describe an individual centrality with respect to the universe – that one is a star at its center. Moreover, this centrality cojoined with directionality implied a freedom to direct oneself and move in any direction with the aim of striking some balance across the universe (Sturgis, 1999). The Iroquois Haudenosaunee nations thought of freedom so highly that they codified notions of individual liberty into their *Great Law of Peace* which foredated the United States Bill of Rights by many centuries. On this point, Cadwallader Colden, a political scholar from the 18th century wrote, "The Five Nations [of the Iroquois] have such absolute notions of liberty that they allow no kind of superiority of one over another and banish all servitude from their territories" (1958, p. xxvii). Although we will examine the virtue of power more in coming sections, it interesting to note the enlightened and longstanding perspectives of Native Americans in this domain. Some tribes would purposefully limit the size of communities such that the power of voice was shared among all members. More fundamentally, some tribes' language itself reflected an egalitarian notion of power across gender. For example, Cherokee men and women could both lead council and if speaking, there were no equivalent gendered pronouns (e.g., he speaks at council and she speaks at council), but rather "a Cherokee speaks at council" (Sturgis, np.). These indigenous peoples knew well that the rules we set, the actions we take, and the language we use can open (or close) space for humans to attain freedom.

By this point, hopefully, you are beginning to imagine the ways that freedom might manifest in your own mathematics classroom. Su (2020) describes three such ways our students might experience a mathematical freedom. The first is a *freedom of knowledge*. Returning to the earlier example involving 32 × 25 (see Figure 4.1), my knowledge of multiplication directly shapes how freely I may approach that problem. If I only have the standard algorithm in my toolkit, I am considerably constrained and unable to move freely in this mathematical space. Su writes of a conversation he once had with a friend about the game of chess,

> It is a sign of a skilled chess player the he or she can play well out of position or circumstance on the board. The

> person who is unaware of his or her options is like a player who is in a bad position. Because even if there are fruitful paths available to you that you are not aware of, they may as well not exist.
>
> (p. 169)

In an earlier chapter, I shared an exchange around the task 400-198 and how Malik was initially unwilling to engage. While that example was couched in the virtues of meaning and perseverance, it is quite easy to consider this moment through a lens of freedom as well. Bound to the standard algorithm, Malik presumably contemplated a seemingly endless process of borrowing and carrying and found himself trapped (i.e. "I am so done with this one"). Had he the knowledge of other strategies, Malik would have enjoyed a freedom of movement through that task. Fundamentally, this is one of the primary arguments for developing mathematical knowledge (of any type) in the first place. A rich understanding of mathematics allows us the freedom to engage with the world and its myriad complexities in different ways. Knowledge lends us a choice in the matter.

Another type of freedom described by Su (2020) is the *freedom of exploration.* In the preceding chapter, we examined the virtue of exploration and the fulfillment that comes from acting on our curiosities to seek the unknown. However, undergirding such exploration is a presumption that one is free to do so. Quite often, there are interesting mathematical offshoots and blind alleys our students might like to explore (e.g., why is dividing by 0 undefined and not infinity?), but the constraints of our curriculum or our schedules prevent us from doing so. In a very real sense, our students' freedom to explore can be constrained in many ways. Sometimes these constraints are due to factors beyond our control as teachers. In other instances, though, there are often moments when our hands are controlling the levers of freedom for our students, and we may choose to move toward exploration or away from it.

Rebecca Brown Corwin led a life committed to mathematics education. She earned her doctorate in mathematics education from Harvard University, was a fifth-grade teacher, and wrote numerous books and articles during her time as a university professor. In one such article, Corwin (1989) recounts her own experience as a child learning mathematics in school. At the onset, she describes herself as a bright finder of mathematical patterns. Corwin continues,

> We had learned that we could check subtraction by adding the subtrahend to the difference and if the answer matched with the original number, we knew we were correct. We could check division by multiplication. In my eagerness to check my own work, I taught myself my own checking method for multiplication. First, I did the problem 'downwards'. This meant multiplying by the ones digit in the multiplier first and then by the tens digit, the usual method . . . Checking it simply meant multiplying 'upwards', since the distributive and commutative properties exist. (Of course, no one in those olden days talked about the 'distributive property' or 'commutative law' For all practical purposes, I invented them or somehow assumed them) . . . You can imagine the satisfaction with which I realized that I could actually do multiplication problems in two ways, and that I could always get the same answer even though the intermediate steps were different.
>
> (pp. 223–224)

Let's pause for a moment and examine Corwin's mathematical thinking. While Corwin provides some more detailed description of her "upwards" method, perhaps considering this method in the context of a problem is more expedient (see Figure 4.2). Take a moment to think about how Corwin's method differs from the traditional algorithm for multiplication. Try to discern how that "upwards" method works.

Perhaps, upon seeing these two strategies aside one another, you had some insight about Corwin's new way of solving multiplication problems. Perhaps you are still thinking about this. Thinking through the mathematical inventions of our students can be challenging but also quite fun. Are they mathematically valid? Do they work for other, similar problems? In this instance,

Traditional "Downwards" method	Corwin's "Upwards" method
32	32
x 24	x 24
128	48
640	720
768	768

FIGURE 4.2 Corwin's Upward Method

I would draw attention to the partial products in the traditional "downwards" method. One hundred twenty-eight is the product of 32 × 4 and 640 is the product of 32 × 20. Now, consider the partial products of Corwin's "upwards" method. I will stop here to avoid spoiling your mathematical experience:

How does Corwin's method work? Is it mathematically valid? Why or why not? Jot down your notes here. Share them with your students. See if they can build on your ideas.

Returning to Corwin's essay, she shares of how, upon moving to a new community, her mathematical experience changed drastically. Corwin writes,

> Mrs. Arsenault sent us to the board to do multiplication problems . . . I did a four-digit by two-digit problem upside down [i.e., using the upwards method] and stood next to it waiting for praise. I was nervous. This was the

> first time I'd been to the board in the new classroom. Mrs. Arsenault looked at my work and then down at the teacher's manual in her hands. Up again to the board, and down to the book. My heart began to race and my palms got sweaty. What was happening? I was always right, and I was sure I was right this time . . . "What have you done" [Mrs. Arsenault] asked very angrily. I was terrified, but I explained my 'special' method. *"How can you do it like that?"* she asked. I misinterpreted this as a request for information and explained again, including the information about how it always worked no matter which way it was done. "I have never seen anything like that", she thundered. "Erase it. It's wrong. Never, ever do multiplication like that in this room again." I was horrified. I couldn't do anything but erase the offending numerals shakily and try to get to my seat without wetting my pants or crying. I sat in shame for days, fearful of doing anything to deviate from her rules.
>
> (pp. 224–225)

If you are like me, then you probably had a quite visceral, negative reaction to reading that. There are so many missed opportunities for Mrs. Arsenault to engage with Corwin and her mathematical ideas. The hurt and damage done in this exchange is so needless. When I first read this essay, I found myself very angry with Mrs. Arsenault and the choices she made that day. With the passing of time, that anger became more akin to sadness. I was sad that Mrs. Arsenault (as she returned repeatedly to her teacher's manual) did not, apparently, have the freedom of knowledge to engage with Corwin's thinking. This limitation, in turn, directly shaped Corwin's freedom in that mathematical space and quite drastically. This is to say nothing of the effects such moments have on our students' sense of self-efficacy and mathematical identities. Of course, this is a fairly extreme example, and if you are reading this book, my guess is that you would likely react quite differently from Mrs. Arsenault. However, on the topic of freedom (and freedom to explore in particular), if we are critical of Mrs. Arsenault's approach, then we are obligated to create a more liberated mathematics classroom. We are obligated to build a classroom culture where students can freely share their ideas, their upwards methods, their sideways methods, and so forth. Imagine if, instead of saying "Erase it, it's wrong", Mrs. Arsenault had said, "Rebecca, that's interesting. I have never seen that before. If you are comfortable, talk us through how it works. Does it work with other problems?" Let freedom ring!

The Virtue of Power

It is nearly impossible to have a discussion of freedom without considering power. I have the freedom to walk into one of the many luxury boutiques on 5th Avenue in New York City or Rodeo Drive in Beverly Hills, California and buy anything I would like. However, the power afforded to me by my current bank account balance limits me considerably. If freedom is the terrain, power is the vehicle we use to move about. Therefore, exploring the virtues of freedom and power in concert makes sense. But wait, is power really a virtue? Like other virtues, it depends upon how it is wielded. In the same way that curiosity can take on a morbid form, or unsupervised play can result in permanent marker drawings all over my basement walls, power can be used for more negative ends. Conversely, when we contemplate fulfillment and human flourishing, power is quite like other virtues discussed in this book – to be *empowered* is, quite literally, to have power. As teachers in the mathematics classroom, the default condition is for power to centralize with us, but how might we lend some of that power to our students so that they can realize their own freedoms? When and how much power should we share? Before diving too deep into these challenging questions, let's examine what we mean, exactly, by power.

Power, in its basic form, is the capacity to manifest one's will even in the face of challenge. Quite simply, power is the ability to make my intentions real. If, in a karate class, my intention is to break the wooden board in front of me, power (channeled through my fist or foot) is what allows me to do so. Another term we might layer into the mix is *authority.* In many instances, we might use power and authority interchangeably, but I would like to draw a careful distinction. In his discussion of authority and power, Barnes (1984) writes,

> the received view of authority within the sociological tradition is that it is power plus: power plus consent, or power plus legitimacy, or power plus institutionalism . . . power holders, it is said, do well to seek legitimacy and consent as soon as they may, so that they can enjoy the enduring benefits which flow to the occupants of authority.
>
> (p. 180)

As teachers, we retain not only power, but legitimized authority that comes with being the adult in the room who is charged by our district to lead the learning experience. This is noteworthy because, in our mathematics classrooms, power alone may not be sufficient to realize one's freedom. Rather, as teachers, we lend our authority to students as well, and we can think of this as power legitimized. In the earlier example, Rebecca Corwin had the mathematical power to create a novel way of doing multiplication, her "upwards method"; however, Mrs. Arsenault did not legitimize this power in the mathematics classroom (i.e., "Never, ever do multiplication like that in this room again".). Thus, when we think about how to share power with our students, it is important to remember that we are also sharing some of our authority. We are authorizing our students to be empowered. Think of it as our stamp of approval which legitimizes the power of our students.

At this point, you might be wondering about how much power we, as teachers, should retain. Indeed, there is a large distance between Mrs. Arsenault and maintaining a productive and orderly classroom culture. Returning to Rousseau's notion of freedom within agreed-upon guardrails, teachers should not extend to their students' limitless power, nor should we allow our students to assume power that has not been granted or legitimized. A few years ago, I was leading a summer camp session for young children at my university. This session, focused on engineering different types of flying vehicles (e.g., fancy paper airplanes), was held in a science lab. As my campers entered the lab, one kiddo (about seven years old, I am guessing), immediately unbent a paperclip and inserted the wire end into an electrical outlet. Thankfully, the incident ended without harm, albeit with plenty of sparks. In that moment, the camper perceived freedom and seized the power to explore the intersection of power outlet and paper clip. However, this was not a legitimate expression of power to realize freedom. As teachers, it is our responsibility to be judicious in the power we lend and legitimize. It's not a free-for-all to empower students to do anything that pops into their minds. My point, rather, is that legitimate power in the mathematics classroom originates with the teacher, and it is up to us to share and distribute that power in productive ways so that our students may experience a freedom that is not absolute but fulfilling.

If we consider human flourishing (i.e., eudaimonia) as the goal of virtuous pursuits, then legitimate power most certainly assumes the mantle of virtue itself. Perhaps you find this assertion a bit disconcerting. Initially, I did as well. For much of my life, I was conditioned to view power skeptically. Power corrupts and absolute power corrupts absolutely, right? However, power also provides an essential key to unlocking fulfillment. In addition to serving as the mechanism for the realization of freedom, power also connects to Aristotelian virtues in that empowerment comes "with a moral autonomy that enables self-discipline, self-monitoring, and self-control" (Shatilova, 2024, p. 59). Quite simply, to be legitimately empowered is to experience an autonomy that allows for cascading virtuous dividends as one learns to wield agency in fulfilling ways.

Turning to power in mathematical spaces, we might consider two variants put forth by Crouch (2013), *stuff-making* and *sense-making*. Su (2020) elaborates on these forms of power:

> *Stuff-making* refers to not just to potency of humans, but the potency of the world. Electric power as a form of energy is used to make stuff. People make stuff and change their surroundings. Mathematicians have a nice word for this: *transformation*. Just as mathematical functions transform the things they operate on; creatures of the world are transforming their environments. The universe is in a

> common state of transformation . . . *Sense-making* describes the ability to understand the world and, in its fullest expression, to make meaning of the world. It's also creative – it requires imagination. People make sense of the world . . . Math explorers do both things. We make stuff – we make definitions, we create structures, we prove theorems, we develop model – but we also make sense – the models and symbols we create are invested in meaning.
>
> (pp. 130–131)

What these authors describe are different types of power our students might wield in a mathematical environment. When our students are legitimately empowered, they can create, they can build, they can make stuff. When empowered, they can also make sense of their activities, they can generate meaning and relevance for what they are doing. In a very real way, legitimate power and freedom are the keys to realizing many of the other virtues in this book.

Free Association: Let's Draw Something!

Draw a picture of something free or something powerful (or both). It can be a detailed work of art or a stick-figure sketch, take as much time as you would like. Once you are done, how might math connect to what you have drawn?

Make your sketch here. Share it with your students. Ask them to make sketches of their own and look for math connections

Building a Culture of Freedom and Empowerment

At first blush, I suspect many teachers hear the words *freedom* and *empowerment* and think of them as good things to aim for in a classroom. However, in the often tightly packed confines of an actual math lesson, it can often be challenging to imagine how to achieve such goals. A teacher might say, "I love the idea of my students being free and empowered to follow their math ideas in all sorts of directions, but they will be tested on standards, and I have to teach in that direction". Perhaps another might chime in, "and might all that freedom and power just lead to confusion among the group?" While I am sympathetic to these perspectives, I would argue that attending to curriculum, standards, and assessment is not mutually exclusive to creating a classroom culture of mathematical freedom and empowerment. These ideas can and do co-exist in many classrooms when we think about degrees and apportionment. As mentioned earlier, what we are talking about here is not the extension of freedom and power without limits to our students, but rather a sharing experience such that we, as a classroom community, may partake of these virtues cooperatively.

Returning to Rebecca Corwin and her upward way of doing multiplication, it would have cost Mrs. Arsenault nothing, other than a bit of time, to invite Rebecca to share more about her thinking – to open up space for Rebecca's freedom of thought and empowered her to share more about her ideas. That's not to minimize the time involved in these efforts. Building a culture of freedom and empowerment (as with building for any virtuous culture in the mathematics classroom) takes time. It is a deliberate choice, and the choices we make in-the-moment are, inherently, steeped in opportunity cost. We choose to do A at the expense of B. This is what Williams (1961) referred to as a *selective tradition* where a person's choice (the teacher, in this instance) creates space or legitimizes one vision of learning at the expense of another. My contention is that we may pick our spots strategically and thoughtfully to lend freedom and share power such that classroom culture itself takes on a new form. Indeed, this will cost some time, but what we gain from that expense is new environment that is far more mathematically propulsive.

So, what exactly can we do to create this classroom culture? Lots of things, I say. Let's begin by thinking about the core practice of questioning. A key element of the mathematics classroom is questioning. In an earlier chapter, we examined the role of questioning through a lens of meaning and perseverance – remember Reinhart's (2000) titular guidance to "never say anything a kid can say". The practice of questioning also has connections to the virtuous pursuits of freedom and empowerment, too. In an interview with *Edutopia,* José Vilson, a prominent middle-level math teacher and scholar, talked about how good mathematics education is a civil right. Embedded in that highly illuminating conversation, he described his view of questioning as an empowering stance. Vilson remarks,

> When students get stuck on a problem, I ask them a series of questions to try and figure out where they're having trouble. Then I try to scaffold things for them. But at the precipice, when they're just about to get it, I say, 'OK, you got it from here. I'm not going to tell you yes or no.' Then I walk away.
>
> (Boryga, 2024)

What Vilson describes is a stance intended to build mathematical confidence not through explanation but questioning. Fundamentally, this is an expression of productively shared power. Indeed, in this example, the teacher is expressing power by formulating and posing questions. The student is extended power in the form of a response, and in Vilson's practice, the student's response is something of a final word on the matter, at least for that moment. The key, here, though, is the type of question that we pose. If our questioning approach is received as a forced inquisition, that is not commensurate with shared power. Too often, questions in the mathematics classroom are fairly closed, forcing a dichotomous response (yes/no), or an attempt to lead the student to verbalize a correct answer. This is what Shelly Harkness (my dear mentor) and I refer to as behavior eliciting

(Thomas & Harkness, 2016). Think of this as the instructional equivalent of leading a witness in court. The purpose of questioning, in this form, is merely to guide the respondent to say some intended words. More open forms of questioning, though, are a different story entirely. *How did you think about that problem? What do you think we should do next? Why? What does this problem make you think of?* These and similar questions are far more productive. They lend the student's thinking legitimate power and create space where that individual can freely share. Astute readers will note that these are the same sorts of questions we discussed in our examination of meaning and perseverance. Asking how one thinks about a problem not only lends power and freedom, but that question also creates opportunity for meaning making. This is a wonderful example of how virtues may intertwine and stack in a mathematical moment to enhance human flourishing.

Another way to think about freedom and power in our classrooms relates to mathematics tasks. At risk of being overly simple, we can think about the freedom and power students realize *within tasks* and students' freedom and power to *shape tasks.* Let's begin by thinking about the freedom we might enjoy within a task. Consider the task below (see Figure 4.3). Working with pattern blocks in this manner is fairly common and one can find examples in many school curricula (e.g., Illustrative Mathematics, 2025).

This task is an example of a non-routine, or *divergent,* task that we explored in a previous chapter. Similar to the way questioning is conducive to meaning-making and power sharing, these sorts of open-ended tasks are similarly supportive of meaning as well as freedom in our mathematical spaces. For this task, we have specified that a triangle is to be created and, embedded in that specification, are some core mathematical ideas (i.e., a polygon with three sides and three angles), but beyond that, students are *free* to experiment with the different blocks – see which ones might fit together to create the desired shape and which one's don't. This freedom comes from the many different ways that students might negotiate this task and achieve the goal. Tasks that are more open-ended, non-routine, or divergent, necessarily

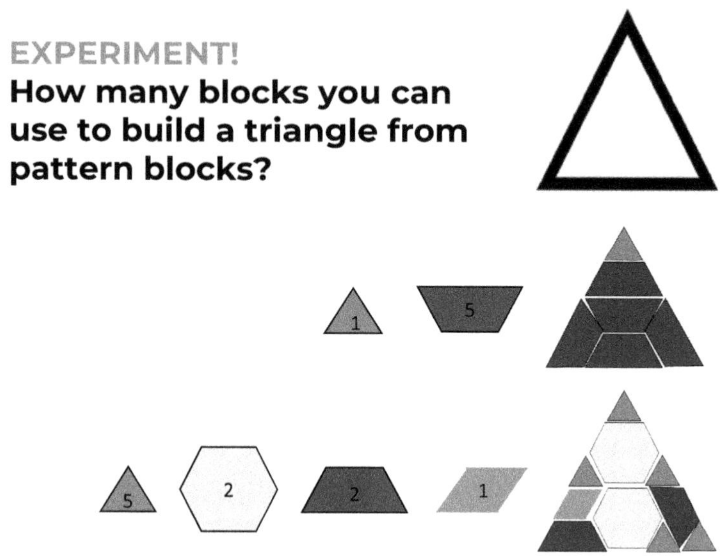

FIGURE 4.3 Pattern Block Triangles

create spaces for freedom within and lend students power to exercise that freedom.

What about the shaping of tasks themselves? When and how might we create liberated spaces and lend students power to create their own mathematical tasks or influence task design in some way? This can go beyond merely allowing students to, say, choose the numbers for a subtraction problem. Station teaching, sometimes referred to as center-based teaching, is one way to afford students freedom and power in the types of mathematics tasks that they experience (Andreasen & Hunt, 2012). While mathematics stations or centers themselves can and should be carefully designed to maximize learning opportunities, there is also opportunity to empower students as they move about and direct their own experiences at some distance from the teacher. Classroom conversations around favorite types of tasks and task contexts and then building these ideas into subsequent lessons, are also ways to lend students power to shape their learning experiences. Giving students, on occasion, the autonomy to design a task or choose which manipulatives to work with for a given period is creating freedom and sharing power. Of course, as expert teachers, we want to choose our spots carefully. Often, we design tasks with very specific aims in mind. However, there are also opportunities in most any math lesson where we might check in with our students. "What tool would you most like me to get out of the math closet for this task?" "What do you think would be a really good problem we could work on that gets at this math idea?" "What real-life setting do you think is a good fit for this problem? Why?" These and similar questions allow for the sharing of power and the enlargement of freedom in the classroom.

Core Practice: Mindful Managing of Power

When we consider the interactions of our classrooms, teaching in this manner requires us to carefully manage power. In every interactive moment (and I hope there are many in your classroom), there is an intersection of power. Sometimes that intersection is at obviously differing levels (i.e., teacher and student interactions). Sometimes the power differential is more subtle, such as when students interact with one another. Rarely, though, is power evenly distributed among those involved. To maximize individual students' realization of freedom and their own mathematical empowerment, we must manage the complex power network in our classroom communities.

The first step in this power management process is simply becoming aware of its distribution. Again, as teachers, we retain much of the power in a classroom by default, but there are still complicated power dynamics among our students that are quite fluid. When you group students, who tends to lead or perhaps even monopolize the conversation? Who appears left out? Which students are always vying to contribute to a discussion with anxiously raised hands? Which are quieter? Perhaps some students are really deep thinkers, but hesitant to share. Perhaps others really open up in small groups or pair activities, but only with certain other people. Perhaps you may notice some patterns here regarding gender or marginalized groups in the manifestation of power. Key, though, is to diagnose how power is moving through your students, where it flows, and where it ebbs.

Once we have a sense of power movement and potential imbalances, we can begin actively apportioning and rebalancing. Assigning group roles to minimize individual dominance, strategic grouping and pairing for maximizing interactions, using talk moves to shift power away from students who may be inclined to dominate discussions and toward others who may be more reserved. Note, these sorts of moves may require some behind-the-scenes work with individual students. Sometimes, it can be discomforting to suddenly experience freedom and empowerment in a space when, in previous classes or earlier grades, you had gotten used to a quieter role. In these instances, teachers might connect with students in 1-1 settings to gauge their willingness to lean into a moment that opens to their perspectives, their ideas, and their voice. Perhaps we give some affirmation of their value in the community and some encouragement about their worthiness to assume power and exercise freedom. All the while, we are mindful of our own power and how we wield it in the classroom. When is it important for me to exert some influence over a task design, group discussion prompts, or assessment? When might I lend power in some of these spaces for my students to find their own paths? Unfortunately, there isn't a fixed series of steps or processes we can walk through here, partly because so many of our decisions depend upon the different students who make up our community. Each individual brings with them a lived history and personality that influences how power will flow and freedom will manifest. Our role, in this domain, is to be mindful of these flows and considerate of how power is being apportioned. The idea that we will ever achieve some perfect balance of power is almost certainly impossible, but what we can do is think of ourselves as active managers of a power portfolio. Just like a diversified stock portfolio that doesn't favor a single company too much, we can think about who, in our group, is exercising outsized power, which likely is constraining others' freedom of voice. We can actively apportion and reapportion power (including our own) to help all of our students find as much freedom in a moment as possible.

Professional Noticing for Freedom and Power

Returning to our connecting practice of professional noticing, recall that this practice is comprised of three components – attending, interpreting, and deciding (Jacobs, Lamb, & Philipp, 2010). When we attend, we want to observe closely to gather details. From there, we interpret that moment and all to form some impression of events (e.g., students' understanding, and group dynamics), and then, with that interpretation in mind, we arrive at a decision. The key is that our instructional decisions flow naturally from our interpretations which, themselves, are a direct product of what we are seeing and hearing. As with other virtues, we can leverage this practice to closely examine our instructional moments in terms of freedom and power. Consider the following exchange around a cupcake task (see Figure 4.4).

If we had 3 different icing types, vanilla, chocolate, and strawberry. And we had 3 different cake types, chocolate, vanilla, and confetti sprinkles . . .

How many different types of cupcakes could we make?

TEACHER: *"I had a cupcake last night, but it wasn't my favorite. My favorite is probably a vanilla cake with strawberry icing. So, how many different types can we make?"*
SHAN: *"I like all chocolate, so that's what I would make.*
TEACHER: *"I feel that, but what about different types? How many different versions could you make? See if you can draw what those versions would look like."*
SHAN: *"Three"*
AMARI: *"No, nine, there's too many ingredients for just three. We have to draw nine." [begins drawing on chart-paper]*
TEACHER: *"Hold up, Amari, for just a second. Shan, tell me why three."*
SHAN: *"You just put vanilla with chocolate cake, chocolate with vanilla cake, and strawberry with confetti cake.."*
TEACHER: *"Are there any others? What if we put strawberry with vanilla cake, like my favorite?"*
AMARI: *"That's what I am saying. They can all go with each other." [begins drawing again]*
TEACHER: *"Amari, I want to hear more from Shan. Hold up on the drawing. Shan, what do you think about those different options?"*
SHAN: *"Well if you can move the stuff, then it's like chocolate could go with chocolate."*
TEACHER: *"Your favorite, right? Shan, see how many versions you come up with. Amari, see if Shan's thinking matches with what you are thinking and then you all can get to drawing."*
SHAN: *"What if you had, like, toppings, like sprinkles or a cherry, or coconut. Then you have a bunch more options I bet."*
TEACHER: *"I love it! When you all get done with your drawing, see if you can decide on a topping and then add that in the mix for another drawing. Let me know how that changes the number of options."*

FIGURE 4.4 Cupcake Combinations

First off, let's examine the task. In this instance, we have three possible icing types and three possible cake types and students are asked to determine and represent the total number of different cupcake variants. This task is consistent with categorizing and data representation standards in both kindergarten and first grade (NGA & CCSSO, 2010). As an aside, a version of this task has also appeared on past versions of the National Assessment for Educational Progress for fourth grade (NAEP, n.d.).

Turning to the students' thinking and activity on this task, Shan asserts that there are but three possible combinations. This is a somewhat common misconception that, ostensibly, stems from the manner in which the task itself is presented – one icing type above one cake type. Quickly, Amari corrects Shan and moves to control the drawing. The teacher appears to *attend* to this interaction, *interprets* this exchange as a power imbalance and *decides* to counterbalance this moment by pausing Amari's efforts so that Shan might have more space to process and verbalize the mathematics. This pattern of imbalance manifests later in the moment as Amari, again, seeks to move forward with a drawing as Shan is still considering the impact of pairing different icing and cake types. Again, the teacher seeks to provide a counterbalance in this moment. Experienced teachers will recognize these power moves almost innately. Managing the interpersonal interactions of students is so common to the work of teaching that it seems to move into an almost subconscious space. My point, in this section, is to draw that work back out from the subconscious so that we may consider it consciously. In these complicated mathematical moments, we want to keep our minds tuned to who is wielding power, who is perhaps being overpowered, and what we might do to achieve a productive balance. Again, there is no magic concoction of steps to take in this area. Each moment will be unique in its own ways, but when we, as teachers, become attuned to power and our role as balancers, our capabilities to respond in effective ways grow considerably.

The exchange ends with Shan hypothesizing about the impact that toppings (e.g., sprinkles and cherry) might have on this scenario. In this moment, the teacher elects to expand the task space to accommodate this thinking (i.e. "I love it!"). Or, in other words, he affords students additional freedom within the task space, albeit with a key caveat – that they finish the original task first. This is a good example of how freedom with guardrails might manifest in a mathematical moment.

Questions for Reflection

At the end of each chapter, I offer up some questions that, I hope, inspire some interaction and conversation. These ideas of how freedom and power inflect and inform our mathematics teaching are complex and worthy of contemplation. As in other chapters, the questions I pose below do not have easy answers, but my hope is that they propel you in a productive direction and keep you thinking hard about freedom and power in the math classroom.

- *How do you balance the demands of your curriculum and instructional pacing with opening space for students guide the math activities of a lesson? In what ways and how often do students lead instruction in your classroom?*
- *How do you think about student grouping and pairing with respect to power? Often it can be advantageous to group heterogeneously in terms of mathematical skills and knowledge, but this also introduces power differentials that complicate interactions. How do you think about and resolve these power differences when they occur?*
- *How do you think about your own power in a math moment? When do you feel the need to exert power and when do you feel more comfortable stepping back and letting students claim power?*
- *How much freedom do you experience as a math teacher? Do you feel like you have the freedom to enact the practices described in this chapter? Does your curriculum or school's policies allow you to share power with students and lend them freedom in the math classroom?*

- *How would the parents of your students and other individuals outside your school community think about these ideas of teaching for freedom and empowerment? What might they think of giving students freedom to shape their math learning experiences? If you imagine that they would be critical, how might you respond?*

References

Andreasen, J. B., & Hunt, J. H. (2012). Math stations for commonsense inclusiveness. *Teaching Children Mathematics*, *19*(4), 238–246.

Barnes, B. (1984). On authority and its relationship to power. *The Sociological Review*, *32*(1), 180–195.

Boryga, A. (2024, May 10). *José Vilson: Good math education is a "civil right."* Edutopia.

Cadwallader Colden. (1958). *The history of the five Indian nations depending on the province of New-York in America*. Ithaca, NY: Cornell University Pres

Corwin, R. B. (1989). Multiplication as original sin. *Journal of Mathematical Behavior*, *8*, 223–225.

Crouch, A. (2013). *Playing god: Redeeming the gift of power*. Downers Grove, IL; InterVarsity Press.

Eckstein, W. (1944). Rousseau and Spinoza: Their political theories and their conception of ethical freedom. *Journal of the History of Ideas*, 259–291.

Hansen, M. H. (2013). *Reflections of Aristotle's 'Politics'*. Copenhagen: Museum Tusculanum Press.

Illustrative Mathematics. (2025). *IM-K5 curriculum*. Dubuque, IA: Kendall Hunt.

Jacobs, V. R., Lamb, L. L., & Philipp, R. A. (2010). Professional noticing of children's mathematical thinking. *Journal for Research in Mathematics Education*, *41*(2), 169–202.

Johansen, B. E. (1982). *Forgotten founders: How the American Indian helped shape democracy*. Boston, MA: Harvard Common Press.

NAEP questions tool - ETS. The Nation's Report Card. (n.d.). https://www.nationsreportcard.gov/nqt/

National Governors Association (NGA) Center for Best Practices & Council of Chief State School Officers (CCSSO). (2010). *Common core state standards for mathematics*. Washington, DC: Authors.

Ni, P. (1993). The Taoist concept of freedom. *Grand Valley Review*, *9*(1), 23–30.

Reinhart, S. C. (2000). Never say anything a kid can say. *Mathematics Teaching in the Middle School*, *5*(8), 478–483.

Scott, J. T. (1997). Rousseau and the melodious language of freedom. *The Journal of Politics*, *59*(3), 803–829.

Shatilova, I. (2024). *A virtue ethics approach to the ethicality of employee empowerment* (743). [Doctoral Dissertation, Jyvaskyla University]. JYU.FI.

Simpson, M. (2006). *Rousseau's theory of freedom*. New York: Continuum.

Sturgis, A. (1999). Liberty in perfection: Freedom in native American thought. *Foundation for Economic Education*. Retrieved from https://fee.org/articles/liberty-in-perfection-freedom-in-native-american-thought/

Su, F. (2020). *Mathematics for human flourishing*. New Haven, CT: Yale University Press.

Thomas, J., & Harkness, S. S. (2016). Patterns of non-verbal social interaction within intensive mathematics intervention contexts. *Mathematics Education Research Journal*, *28*, 277–302.

Wei, R. C., Darling-Hammond, L., & Adamson, F. (2010). *Professional development in the United States: Trends and challenges*. Dallas, TX: National Staff Development Council.

Williams, R. (1961). *The long revolution*. London: Chatto and Windus.

5

Beauty and Truth

For each chapter, I have curated a playlist from my own music library. These are songs that put me in the mindset of the topics at hand. Some may be on the nose and others may feel like a bit of a stretch. Hopefully, though, you find the music an enjoyable addition to your thinking in these areas.

Note: Spotify automatically adds selections to a given playlist. My original selections are the first 6–8 songs on a given list

https://open.spotify.com/playlist/2Nx0bDjsWWvxYgJFsuf73E?si=6dZ8_Z5uTbSL3vJN35bbjg

Introduction

Several months ago, I was in Spain for work. My university has a program for preservice teachers to complete a portion of their student teaching assignment abroad. To meet certification requirements in our state, however, someone from the students' accredited teacher education program still needs to observe their teaching and verify the strength of their knowledge and skills. Thus, I found myself in Valencia, Spain, for a week, where I spent time in a local elementary school observing my students develop their teaching practices and do some amazing work with children. In addition to this richly rewarding work, I also had

DOI: 10.4324/9781003484783-5

the opportunity to sightsee a bit. Having never been to Spain, my partner and I set about exploring as much as possible in our spare time including a quick visit to Barcelona, which is a little over 200 miles (350 km) north of Valencia along the Mediterranean coast.

In our short time in Barcelona, we visited one of the most famous churches in the city, the Sagrada Familia (see Figure 5.1). Interestingly, the Sagrada Familia is the largest unfinished Catholic Church in the world. Construction began in 1882 and continues to this day. Designed by Antoni Gaudí, an architect and designer of renowned fame whose fusion of modernist and gothic styles was influential around the world, this church (a basilica, actually) features an astounding array of ornate windows, towers, and spires that boggled my mind as I tried to take it all in. Everywhere I looked, I saw overwhelming beauty and grandeur in Gaudí's masterpiece.

An interesting fact about Gaudí is that he was skilled in many actual building techniques. Gaudí had experience shaping iron, creating stained glass, sculpting with ceramics and plaster, and so forth. His intimate knowledge of skilled tradecraft allowed

FIGURE 5.1 Sagrada Familia Exterior and Interior

him to push the boundaries of the possible in his architecture and designs. This also led to a deeply held sense of integration in his projects. Gaudí worked from a notion of connectedness among all elements of the human experience from the spiritual to the physical including, even, the *mathematical*.

As my partner and I walked around the exterior of this massive structure, I looked up and noticed an interesting sculptural element on one façade (see Figure 5.2). This element is located on the *Passion Facade* which features elements of the death and resurrection of Jesus central to the Christian faith. Take a moment and see what you notice in this numeral display. Do any patterns emerge?

First off, if you are familiar with magic squares, you might see something recognizable here. However, a typical magic square abides by some rules (Andrews, 1917). For example, a square

FIGURE 5.2 Magic Square of Sagrada Familia

2	7	6
9	5	1
4	3	8

FIGURE 5.3 3 × 3 Magic Square

with a side-length n would include the integers 1, 2, . . . n^2. Thus, for a 3 × 3 square, would feature the numbers 1–9 (see Figure 5.3).

What makes these squares *magic* is that when you add the rows, columns, and even the diagonals, you arrive at a constant sum – the magic constant. In the example above, the magic constant is 15. Magic squares are thought to date back to 2200 BCE (Pickover, 2009); however, the element from Sagrada Familia is obviously much newer. Moreover, it looks like a magic square but doesn't seem to play by the rules. That is, a 4 × 4 magic square should have the numbers 1–16, but this one is missing the 12 and the 16. It also has 10 and 14 twice. We might think of this as a *trivial* magic square in that it uses some numbers more than once and others not at all. However, the nature of the magic square at the Sagrada Familia was not at all trivial to Gaudí. By this point, you have probably deduced that the magic constant of Gaudí's square is 33. In this context, on the Passion façade, 33 holds special import – believed by many to be the age at which Jesus was crucified. In his vision for this monumental achievement in art, architecture, and design, Gaudí sought transcendence by weaving together elements of the spiritual and the mathematical. In Gaudí's eyes, the positioning of this unique table of numbers enhances the beauty of the façade in powerful and unexpected ways. Seeing it in person and contemplating its significance, I agree.

I share this story because, for me, it represents one way that mathematics can serve as a medium for beauty and an amplifier of spiritual truths held deeply by many. Indeed, Gaudí was not alone in his seeking of connectedness within the spheres of design. As inventors of much of modern mathematics, Islamic culture provides many such examples of this sort of mathematical beauty. Medieval Islamic architects and designers created unique *girih tiles* which were a sequence of varying contiguous polygonal shapes creating a tessellating effect on different ancient structures that may be seen on mosques in Iran as well as shrines in India. Lu and Steinhardt (2007) write,

> By laying the tiles edge-to-edge, the decorating lines connect to form a continuous network across the entire tiling. We further show how the girih-tile approach opened the path to creating new types of extraordinarily complex patterns, including a nearly perfect quasi-crystalline Penrose pattern on the Darb-I Imam shrine . . . whose underlying mathematics were not understood for another five centuries in the west.
>
> (p. 1106)

Indeed, these architects and designers from centuries past were also fearless explorers of mathematics and used their profound understanding of geometry to create profound beauty to honor their truths and traditions. Whether it's a magic square, or a tessellating pattern that, when extended, creates "perfect quasi-crystalline tiling" (p. 1107) with infinite order, mathematics has historically been considered across cultures as a key element of beauty and truth-seeking.

In the coming sections, we will explore the virtues of beauty and truth as they relate to mathematics and consider how these ideas might articulate in the elementary mathematics classroom. Indeed, the children in our classes may not yet be ready to examine the mathematics behind complex crystalline tessellations, but the connection of mathematics with notions of beauty and truth is not only attainable for all students, even very young children, but also an important way to elevate this whole enterprise of mathematics teaching and learning into another plane of experience.

The Virtue of Beauty

Whenever I hear the word *beauty*, one of the first things that I think is that *it's in the eye of the beholder* as the saying goes. In other words, beauty is one of those things that we all recognize and define on our own terms. What I find beautiful, another may not, and vice versa. However, philosophers and scholars have given much consideration to beauty over millennia and tried to add clarity and definition to the unique human phenomena of experiencing beauty. This area of philosophy is often referred to as *aesthetics* and influential thinkers from Confucius to Kant have committed to meaningful work in this area. Indeed, Confucius wrote extensively of aesthetic beauty (*Mei*, 美) and often drew explicit connections between beauty and human expressions including music (*Yue* 樂) which hold deep links to happiness (Fu & Wang, 2015).

On this point, scholars note deep links between beauty and goodness within Confucianism. Lang (1991) writes,

> Confucius underlines the unity of beauty and goodness, and he demands that arts should contain morals . . . Namely, beauty is form and goodness, content. Or in other words, the form of arts should be beautiful, but the content of arts, good.
>
> (p. 68)

Japanese Zen philosophers have also explored the concept of beauty, and centuries ago, wrote of *wabi sabi* – the practice of finding beauty in the unconventional, imperfect, and incomplete. Maheux (2016) writes,

> We might say that wabi sabi is concerned with a range of experiences qualified with adjectives such as: earthy, irregular, imperfect, textured, intuitive, relative, ambiguous, contradictory, and so on. Wabi sabi engages us to appreciate such attributes, finding beauty in the contemplation of objects in which we recognize them.
>
> (p. 176)

Taken together, these conceptualizations lead us to a positioning of beauty as something that converges with goodness and may emerge from the imperfect and unfinished day-to-day experiences, including those of the mathematics classroom.

Maheux continues to describe many, different ways we, as teachers, may find children's mathematical activity beautiful as exemplar of wabi sabi principles as they seek, albeit tentatively and imperfectly, some mathematical truth – that we may see beauty in students' seeking and make that beauty explicit to them as they think and act.

Building from the notion of beauty as sometimes transitory but inherently good, Su (2020) describes four different types of beauty pertinent to mathematical thinking – sensory, wondrous, insightful, and transcendent. Sensory beauty refers to the "beauty of patterned objects that you experience with the senses: sight, touch, and sounds" (Su, 2020, p. 71). We may experience sensory beauty as we move gaze upon a mountain vista or listen to our favorite song. Wondrous beauty refers to

> feeling awe, at seeing something amazing, but also in the sense of feeling curiosity – that causes the mind to wonder, to be curious, to ask questions. While sensory beauty usually concerns itself with physical objects, wondrous mathematical beauty always invites a dialogue with ideas.
>
> (p. 73)

Insightful beauty, Su writes, "concerns itself with reasoning. Logically correct deductions are not sufficient for the math explorer. She often looks for the best proofs, the simplest, or most insightful. Math explorers have a special word for this: *elegance*" (p. 75). Lastly, transcendent beauty "arises when one moves from the beauty of a specific object, idea, or reasoning to a greater truth of some kind – perhaps an insight that reveals a deep significance, or a deep connection to other known ideas" (p. 79). Children may experience transcendent beauty when they develop and realize the ability to construct and use composite units. Indeed, moving from a mathematical landscape where everything must be considered in terms of single units (i.e. "ones") to one where a child may use a composite unit to enact a counting-on strategy to solve an arithmetic task is very powerful. At this point, astute readers are likely noting connections between these variants of beauty and other virtuous ideas (e.g., curiosity and exploration) presented in this book.

In this sense, fulfilling experiences in the mathematics classroom may operate as a network with many different virtues colliding and connecting with one another. In this case, we see how experience, exploration, curiosity, and even imperfect activity may be seen and appreciate as beautiful by teachers and students alike.

Let's take a moment and consider how beauty might intersect with a mathematical moment in practical terms. Consider the exchange between a teacher and a kindergarten student, Lara, below (see Figure 5.4). In this instance, the teacher is working with Lara on comparing two and three-dimensional shapes and using informal language to describe their similarities and differences. This mathematical goal is typical of standards at that grade level (NGA & CCSSO, 2010 – see standard K.G.B4).

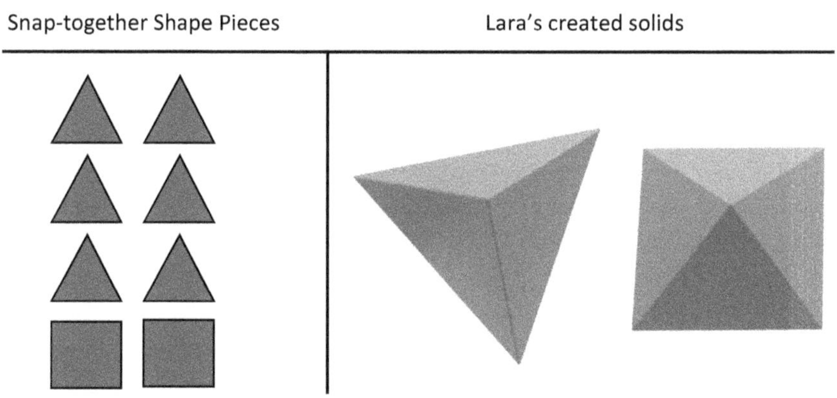

TEACHER: *"So tell me about what you have made here, Lara."*
LARA: *"I made two different triangles. They're kinda the same, but different."*
TEACHER: *"Triangles, like this triangle?" [points to a two-dimensional triangle piece]*
LARA: *"No, the ones I made stand up"*
TEACHER: *"So how are they the same? How are they different?"*
LARA: *"Well, this one [touches the triangular pyramid] is . . .I liked making them, but I think this one is the best."*
TEACHER: *"What do you like about that one?"*
LARA: *"I like the way it sits. The other one [touches the square pyramid] looks like all the rest. This one looks special."*
TEACHER: *"Special how?"*
LARA: *"Like it could make a rainbow if you held it up to a window."*
TEACHER: *"I love it when that happens. You don't think the other one would do that."*
LARA: *"Maybe, but it's like the more normal one. This one [touches the triangular pyramid] is special. I wish it was green . . . like clear, but green."*
TEACHER: *"You know, that one has a special name, tetrahedron"*
LARA: *"tetrahedron [slowly] . . .that sounds like a magic word."*

FIGURE 5.4 Lara's Pyramids

In this exchange, the student has constructed two different three-dimensional figures using square and triangle snap-together tiles. One is a square pyramid and the other a triangular pyramid (or tetrahedron). After noting the distinction between the two-dimensional tile (triangle) and the three-dimensional construction (pyramids), the teacher questions Lara about how she sees these figures in terms of their similarities and differences. Rather than describe some geometric difference (albeit in informal language), Lara chooses to approach her constructions on aesthetic terms ("This one looks special"; "Like it could make a rainbow"). In a very real way, Lara is perceiving beauty in what she has created and sharing that with her teacher. Interestingly, rather than steer directly back into the geometry topic at hand, the teacher engages her for a few moments ("I love it when that happens") and offers up some additional information ("You know, this one has a special name") that, in this case, adds a little magic to the moment. I imagine some of you reading this might be thinking that it's hard to find the time to indulge our students in every possible venture away from the content at hand. I agree. The pacing and scope of typical elementary curricula makes this work quite challenging. However, I would argue that it is this sort of moment (along with others described in this book), that make all of the mathematical content worth pursuing. Why take the time to think about similarities and differences among geometric solids if we first don't find something fundamentally interesting about them? Why not endeavor to begin with some human connection, beauty in this instance, and build outward rather than consider such virtues supplemental to the moment – if time allows.

Interestingly, ancient Greeks also found deep beauty in what were referred to as *Platonic solids* which are multifaceted three-dimensional objects whose faces are all polygons of identical side length. The tetrahedron is a platonic solid while the square pyramid is not. In addition to the tetrahedron, there are only four other Platonic solids – cube, octahedron, dodecahedron, and icosahedron. Fun fact, geodesic domes (e.g., the *Spaceship Earth* structure at EPCOT center in Walt Disney World, Florida) are composed of repeating Platonic solids to approximate a spherical construction. Pickover (2009) writes,

> Plato described the five platonic solids in *Timeaues* in around 350 B.C. He was not only awestruck by their beauty and symmetry, but he also believed that the shapes described the structures of the four basic elements thought to compose the cosmos. In particular, the tetrahedron was the shape that represented fire, perhaps because of the polyhedron's sharp edges. The octahedron was air. Water was made up of icosahedrons which are smoother than the other Platonic solids. Earth consisted of cubes, which look sturdy and solid. Plato decided that God used the dodecahedron for arranging the constellations in the heavens.
>
> (p. 50)

Su (2020) might say that Plato, as he contemplated these solids, was experiencing sensory beauty in their physical nature and appearance and also wondrous beauty as the experience caused his mind to reach for explanations around these things so unique and special. In the exchange between Lara and her teacher, Lara, in her way, is experiencing these ideas as Plato did – "I like the way [the tetrahedron] sits . . .It looks special . . . like it could make a rainbow". Indeed, when we see the possibility for beauty to wash over a mathematical moment, let's do our best to pause and let it happen.

The Virtue of Truth

Some readers of this chapter may have wondered whether something specific guided me to pair the virtues of beauty and truth. Indeed, there is a poem special to me (and perhaps to some of you) that directly links these ideas – *Ode to a Grecian Urn* by John Keats (1897, p. 236). Now in the public domain, a quick internet search will yield the poem in its entirety, but I would like to share the final stanza:

> O Attic shape! Fair attitude! with brede
> Of marble men and maidens overwrought,
> With forest branches and the trodden weed;

Thou, silent form, dost tease us out of thought
As doth eternity: Cold Pastoral!
When old age shall this generation waste,
Thou shalt remain, in midst of other woe
Than ours, a friend to man, to whom thou say'st,
"Beauty is truth, truth beauty,"—that is all
Ye know on earth, and all ye need to know

In this poem, Keats describes a narrator contemplating an ancient urn and all of the historic depictions that are etched into it. Finding it impossible to separate the sensory beauty of the experience from the urn itself, the poem ends with the narrator concluding that beauty and truth are one and the same. My mother was an avid reader of poetry and had many favorites. Some of my earliest memories are of her reading poetry to me while trying to get me to go to sleep. While she had various poetry books and compilations, she also kept a folder of poetry clippings that she would place around the house (fridge, bulletin board, etc.), periodically, of which this was one. Of course, every individual may draw their own meanings, but I have always thought Keats was telling us that whenever we encounter beauty or truth, we can assume the presence of the other. That when we see something beautiful, therein lies some deeper truth about the moment, ourselves, or the world around us. Similarly, when we receive and internalize some truth, there is beauty inherent to that conception and process. In other words, realizing truth is a beautiful act.

Before we get too far ahead of ourselves, though, let's take some time to examine what is meant by truth and how human beings have considered truth over time. In simple terms, I often think of truth as the descriptor of factual knowledge. Expanding on this, Merriam-Webster (n.d.) defines truth as "the body of real things, events, and facts . . . the property of being in accord with fact or reality" but also "a transcendent fundamental or spiritual reality" and "sincerity in action, character, and utterance". These latter definitions signal something deeper going on with this thing we call truth.

One of the great joys of writing this book is exploring the universality of these virtues. Pick nearly any time and any place, and we can find people grappling with these ideas and how they help us understand our existence. Across all nearly all cultures and throughout recorded history, the concept of truth has captivated humanity as we sought to better understand reality and our lived experience – all the things of this world, including ourselves, that were perceived as true. The *Upanisads* are ancient Sanskrit texts, thought to date from as far back as 800 BCE, which form part of a larger body of texts, referred to as the *Veda* (meaning knowledge) that are often cited as the origin of Hinduism (Muller, 2001). In his manuscript titled "On Truth, A Hindu Perspective", Seshagari Rao (1970) writes,

> The Upanisads seek to know the ultimate reality or truth of life from which "this world and our experience have arisen, in which they exist and into which they get merged." They start from an inquiry into the external world and then, step by step, inquire about the body, sense organs, the mind, etc. The quest is existentially important. They involve questions such as "Who am I?" "Whence did I come?" "Where do I go?" "Has life a purpose?" and so on. They ask: "What is beyond the visible body and the invisible mind?" . . . The sages of the Upanisads pursue the quest relentlessly by digging deeper and deeper into the realm of consciousness until they have a vision of the inmost essence of self.
>
> (p. 378)

What these ancient thinkers of millennia past realized was that any contemplation of some truth external necessitates a journey inward to examine why we find it truthful and what that means about us as individuals. This journey, arguably, is one of transcendent beauty as we draw deeper meaning from our sensory experiences. Truth, like beauty, passes fluidly within and beyond us as we move through time and space. Perhaps Keats was onto something when he wrote that "Beauty is truth, and truth beauty" (Keats, 1897, p. 236). More likely, he was grappling with some fundamental idea that captivated the earliest Hindu scholars thousands of years ago.

The Yoruba people emerged as a distinct ethnic group from the populous Middle Niger region of West Africa approximately 4000 BCE (Akintoye, 2010). The Yoruba delved deeply into philosophical ideas including the nature of truth. Oluwatobi (2022) writes,

> In the Yoruba world view for something to be regarded as truth (*ooto*) one thing is needed to be added to complement the empirical evidence and this is the mind (*okan*). The *okan* here is referring to the seat of human consciousness or one's conscience; hence, some refer to it as (*eri okan*) in literary terms, the mind's testifier. Suffices to say that in the Yoruba world view when you claim to know (*mo*) your mind (*okan*) must testify to it before you can claim to lay hold on the truth. The mind so to say, is a mental ability of [humankind], a rational part, and for truth to be fully ascertain among the Yoruba's both empiricism and rationalism have a meeting point.
>
> (p. 5)

What the Yoruba were grappling with thousands of years ago was the tension between the possibility of empirical truths that may exist beyond the individual and the individual's experience of truths – the way that we consciously create and shape our own truths. This individual building of truth comports with the more contemporary *constructivist* perspective of knowledge creation. Simply put, constructivist learning theory posits that knowledge (e.g., the things we know to be true) is not acquired from others but actively constructed by the individual (Piaget, 1970) and in its most radical forms, suggests we may not ever fully know the minds of others as we are unable to access the experiences beyond our own minds (von Glasersfeld, 1995). From this perspective, it is not enough that we, as teachers, attend to something we might think of as an external or universal truth but also to how that truth is being internalized and interpreted by our students. Fundamental to this process is the extent to which we share a conception of reality with those around us.

While challenging in this age of seemingly endless media and commentary bombarding us with competing perspectives and provocative hot takes, the aim of sharing reality, of sharing truth, is deep human desire and one that leads us toward flourishing. Nevertheless, truth-seeking in this time is immensely challenging. Su (2020) writes,

> The world I see today is being upended with political instability stoked by misinformation. Truth is obscure. People are endorsing blatant falsehoods that comport with their worldview rather than accepting complicated truths. We live in filter bubbles that reflect our own biases back at us . . . I hear the hopelessness of my friends' lament "It's just too hard to figure out what's going on so why should I even bother".
>
> (pp. 102–103)

From my perspective, this is all the more reason to lean into the interaction within our mathematics classrooms to maximize every opportunity we may have to share our truths with students and for them to share their truths with us so that we may develop a common understanding of the world around us. Interestingly, in mathematics education, there is a term for this – *taken-as-shared* (Stephan, Cobb, & Gravemeijer, 2003). Taken-as-shared refers to the ways in which individuals may construct a reality that is fundamentally unknowable to others while, simultaneously, sharing some share some common understanding of that reality. In plain language, taken-as-shared describes how we may never fully understand the feeling of, say, sand beneath the toes of another person, but relate, nonetheless. I can never truly know how sand feels on the feet of another, but since I have also experienced sand beneath my toes, we can share something common from that experience. We can talk about the squishy softness or the way it sticks to our feet. We take the experience as something that we share with one another. Truth is similar in the sense that I may never know, precisely, another's conception of an idea, concept or fact, but if I also know something about that same idea or fact, therein lies a sharable experience.

Truth and Beauty Fill in the Blank

Before we move forward with these ideas, let's take a moment to think about how beauty and truth might connect. Below is a table where you may list something you believe to be true and then try to find some beauty in that truth. For example, I stopped at a coffee shop on the way into work (*true statement of fact*), and while I was there waiting for my coffee, like most every day I stop in, I saw several different groups of people beginning their day. When I reflect on what I heard and saw, it strikes me as beautiful representation of the promise of a new day. There is a buzzing energy to the groups that strikes me as hopeful – a sort of caffeinated community beginning again together. Now you try it. What's something true in your world. It could be a simple statement of fact (e.g., stopping by a coffee shop) or some deeply held personal truth. Maybe your truth is math related. Maybe it's not. It's up to you, but the challenge is to find the beauty in that truth, wherever and however it may occur. Ask your students to do something similar. I suspect their revealed truths and experiences of beauty may be quite rewarding.

YOUR TRUTH	*WHAT'S BEAUTIFUL ABOUT IT*
Ex: I stopped at a coffee shop this morning	Ex: There was a buzzing energy among all the people as they started their day. It's like the "newness" of another day expressed in a visible and discernible way.

Building a Culture of Beauty and Truth

Much of what we discuss in this book, pedagogically speaking, boils down to making time for these ideas. However, that doesn't make this common thread any less true or important. While standards, curricula, and assessments all play important roles in the teaching and learning of mathematics, these elements of our classroom landscape remain remarkably mute when it comes to the pursuit of some deeper fulfillment. Quite simply, our instructional resources give much guidance about the what (e.g., content) and the how (e.g., practices) of doing math, but have very little to say about *why* we might invest our time and energy in such a pursuit. In this sense, creating a classroom culture that cultivates beauty and truth begins with us, as teachers, valuing these virtues and committing to create time for their inclusion. Having established our commitment, though, what next?

With time and creativity, there are countless ways we might create opportunities to connect math and beauty. Consider the brief exchange below involving a skip-counting activity (see Figure 5.5).

TEACHER: "*Alright, so let's warm up with a little skip counting. Let's popcorn this, and we will go by threes. I will start us off. Three! Shannon, you're next.*"

SHANNON: *"[whispers "four, five"] Six!, Malachi!"*

MALACHI: *"Nine! Jess!"*

Students continue skip-counting up to 39

TEACHER: "*Ok. . Now we are warmed up. I want to share a poem I wrote and see if you can guess what it is I am writing about. Also, what else do you notice about the poem?*"

TEACHER'S POEM

Yellow, sweet, long
Like it in my breakfast cereal
It gets mushy if it's left out too long
Pull down from the stem to peel before you can eat it.
They grow on trees. I buy them at the grocery. They are with the fruits

FIGURE 5.5 Skip Counting Poems

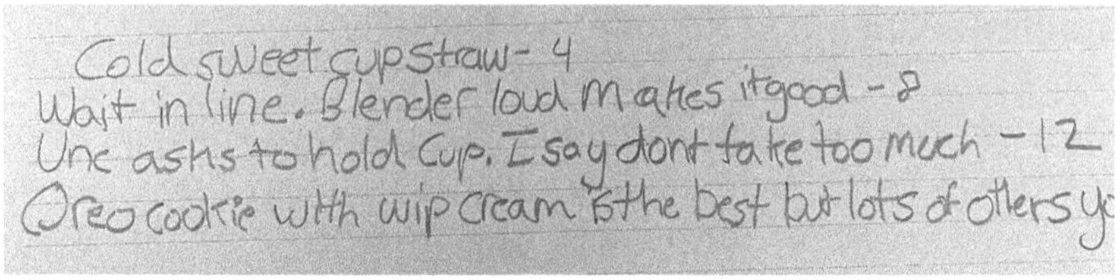

FIGURE 5.6 Pattern Poem

In this moment, the teacher begins with a skip-counting warm-up activity and then presents a poem. Can you guess what the poem is about? Do you notice anything else unique about the poem? Likely, you guessed *banana*, but what else stands out? Perhaps you noticed each line getting longer. Looking more closely (and primed to think about skip-counting), you may have noticed that each line increases by three words (yellow, sweet, long-3; Like it in my breakfast cereal-6, etc.). Now, imagine where you might go with this instructionally? In terms of a mathematical goal, perhaps our aim is to engage in skip-counting or patterning (e.g., 2.NBT.A.2, 3.OA.D.9; NGA & CCSSO, 2010). To pursue that aim, we might ask students to write a patterned poem with the word count of each line increasing by specified amount (e.g., 4, 8, 12, and 16) with the goal of describing their favorite food without giving it away. Imagine what words a second- or third-grade student might use to describe a chicken nugget or a fruit paleta? What about grandma's pierogis? What do you think the student below is writing about (see Figure 5.6)?

Each word that our students might choose relates to some sense memory of their experience with that food – something sensational in the most literal meaning of the word. This is the sort of experience that allows us to connect a mathematical idea with something that holds resonance within us. In the same way that our eyes bathe in the beauty of visual art, our mouths and bodies experience a sensory beauty as we enjoy a favorite food (e.g., cold, sweet, cup, and straw). We can imagine who might be around us as we take that enjoyment and how that enriches the moment. If you don't believe me, try it yourself. Choose a favorite food and then pick a number. Mentally compose a patterned

Mar	**y**	**had**	**a**	**lit**	**tle**	**lamb**	[*rest*]
1	2	3	4	1	2	3	4
lit	**tle**	**lamb**	[*rest*]	**lit**	**tle**	**lamb**	[*rest*]
1	2	3	4	1	2	3	4

FIGURE 5.7 Mary had a Little Lamb

poem about your food and choose your words carefully, savor them, as you compose each line. Now, try it with some students and see what they write.

What about something musical? Music also affords us an opportunity to think about sequencing in a beautiful space. Most music has some type of meter. Simply put, meter is the way that *beats* occur in a particular song (Telesco, 2019). These beats may be simple or quite complex, but what makes music special is that there is some detectable rhythm that guides our listening. A song's time signature, refers to how beats are divided (top number) and the number of beats per measure (bottom number). So, a song with a $\frac{4}{4}$ time signature would have melody structured around 4 beats and contain 4 beats per song measure. The song, *Mary had a Little Lamb* (see Figure 5.7) is an example of a $\frac{4}{4}$ time signature.

Now, where it gets interesting and beautiful is how we enact that time signature. Take a moment to sing *Mary had a Little Lamb* and clap on the 1 and the 3 beats. This produces a particular (arguably traditional) rhythm. Now, try it again and clap on the 2 and the 4 beats. This produces more of a swing rhythm that is detectable to the ear. From jazz halls to gospel choirs, musicians have long understood that hitting a beat on the 2 and 4 is way to accentuate a song in a really engaging way. Once we know something about meters and time signatures, then we can experiment with the music we find beautiful and enjoyable. The next time you are listening to music, clap, snap, or stomp along to it. What rhythm of claps (or snaps or stomps) feels most natural to the song? Try hitting different beats and see if it makes a more pleasing rhythm. Perhaps it makes the song less pleasing.

You probably see where I am going with this – now start skip counting your beats. Choose any sequence you like and see how far you can go, mentally, with your count. Or count out loud if you are feeling bold. Such connections between math (skip counting, in this instance) and beautiful spaces abound should we take the time to look for them.

Now, if you are like me, then perhaps you are a bit concerned about the mathematics getting lost in the poetry or the music (or the artwork, sculpture, etc.). Certainly, my students could gain experience with many more and varied skip-counting sequences if we dispensed with the writing or singing/clapping elements. However, I wonder what we are trading away by choosing that path for every lesson. Cordoning off our lessons from the virtues that make human life worth living, including beauty and truth, ultimately leads to a more fragile and impoverished relationship with math as well as a smaller lens through which we perceive the wonderful things in the world around us.

Core Practice: Slow Down to Smell the Roses

Beginning with the idea of slowing things down in the classroom to create time for virtues, the first step is move more deliberately through the lesson planning process. When thinking about a particular content standard, ask yourself, "How might this idea or concept embody or connect to something beautiful? What deeper truths might this mathematics contain?" For example, working with decimals, students might use 10 × 10 grids and color pencils to create different decimal portions of the entire grid. Depending on their choices, each students' grid would become an individual work of art that contains that represents not only decimals but their aesthetic preferences. Similarly, imagine a fraction task involving giant cookies. Note, I can't take credit for this problem context, rather some amazing research partners developed this one as a way to rehumanize math in the classroom. In this task, there are giant cookies sitting on pedestals in each corner of the room. One pedestal has one

FIGURE 5.8 Giant Cookies on Pedestals

cookie, one pedestal has two cookies, and the last pedestal has three cookies (see Figure 5.8).

Now, task students with choosing which pedestal they would select to get the most share of cookies. You could do this with an individual vote, or setting up stations (i.e., pedestals) to which students might physically move. There is a bit of game-thinking in this task. While the pedestal with three cookies has the most opportunity, it also might draw the most people, resulting in a smaller final share. For example, if 12 people from the class go to the three-cookie pedestal, but only two people go to the one-cookie pedestal, then each member of the latter group receives a larger cookie portion. While this is a good task to examine and compare different fractional quantities, it also opens up space to discuss a deeper truth – fairness. Is this task fair? Does it matter if you are the first person or the last person to choose a pedestal? What is fair, anyway? We will discuss notions of fairness as they relate to justice with more detail in the following chapter.

Now, if these ideas don't come readily to mind, that's completely understandable. Making mathematical connections to beauty and truth (and other virtues) was not an explicit part of my teacher training. However, a wealth of information lives at our fingertips these days. With a few keystrokes, we can gather ideas about songs, poems, and paintings that might connect to a math concept. Perhaps we may even use developing artificial intelligence tools to help us construct something visual or musical that provides us a mathematical springboard into this world. The most rewarding part of these initial explorations into beauty and truth is that they often yield additional material for other instructional moments. We may find something that doesn't quite fit with our planning this week but would be a rewarding centerpiece for another lesson a few weeks from now. Take note of those beautiful things and file them away in your planning materials.

Lastly, we may seek beauty and truth from our students. What sorts of things do they find beautiful? What sounds turn their heads? What visuals draw their gaze? What do they find pleasing? What truths do they hold about the mathematics at hand? On this last, we may be surprised sometimes. Consider base-10 blocks. These are the little cubes (1's), the connected rows (10's), the flat 10 × 10 grids (100's), and the large 10 × 10 × 10 cubes (1000's). Teachers often deploy base 10 blocks to concretize the construction of multidigit quantities and introduce place value concepts and operations. However, students don't always perceive these materials the same way that teachers do – they don't necessarily see the intrinsic "place value-ness" of them. Rather, a student might find these blocks "as disconnected from reality, mysterious, arbitrary, and capricious as the numbers these blocks were supposed to bring to life" (Holt, 1982, p. 219). Or, more succinctly, children see the world through their own eyes and interpret it with their own minds. What clearly is a "ten-rod" to adults, is clearly "one item" to some children. My point is that we must first seek the truths of our students before we can arrive at truths which are taken-as-shared.

Professional Noticing for Beauty and Truth

Revisiting our connecting practice of professional noticing, we want to mindfully apply the three component processes of attending, interpreting, and deciding. In this domain, we want to attend to the emergence of beauty and truth, interpret those manifestations, and enact decisions that simultaneously advance our mathematical goals and honor the beauty and truth of the moment. Certainly, as I discussed in earlier sections, rich instructional planning creates space for beauty and truth, but we must also tune our eyes and ears for these virtues as they bubble up in the students' experiences. Consider the following exchanges around a fraction task involving pattern blocks (see Figure 5.9).

While pattern blocks are a great tool to explore fractional quantities – equivalent fractions and fractional portions of larger wholes. In this instance, note the teachers' creation of opportunities for students to build with beauty in mind ("build something you think is interesting to look at") and directing students to be mindful of that aesthetic dimension ("Be ready to tell me why your pattern is interesting"). Trent and Cara used this opportunity to construct something that held unique aesthetic meaning to them ("Like an angel"; "Like and ancient picture"; "Like a person"). Take a moment and think about what their pattern looks like to you? What do you see in their work?

TEACHER: *"In your pairs, I want you all to use the yellow and red blocks to make a pattern. Any pattern you want, but I want you all to build something you think is interesting to look at. Be ready to tell me why your pattern is interesting."*

...

TRENT: *"We made this one. We started with the yellows and built around it with reds."*
CARA: *"It reminds me of, like, a picture you see on a window in a church. Like an angel."*
TRENT: *"But I think it's older. More like an ancient picture. Not cave ancient, but something early people would make. But it is like a person, sorta."*

FIGURE 5.9 Pattern Block Fraction Task 1

TEACHER: *"I really enjoyed hearing about your patterns. Now, this might be tricky, but I want you all, in your pairs, to try to make another pattern that is one-third of your original pattern. So, whatever you just made, I want you to make something that is one third of that. Take your time, think it through, and for this one, you can use the yellows, reds, blues, and greens if you like. Same thing though, I want your new one third pattern to be interesting and tell me why."*

...

CARA: *"It reminds of those beaded belts or a bracelet something. Like turquoise beads."*
TRENT: *"I think it's more like a futuristic vehicle with green wheels. I still don't think it's right, though."*
TEACHER: *"Why not?"*
TRENT: *"Because it's twenty-seven blocks and if I take one third of that, I mean, divide that by three, which is like one third, it's nine blocks. So, we should have used nine blocks."*
CARA: *"Not all the blocks are the same, though."*

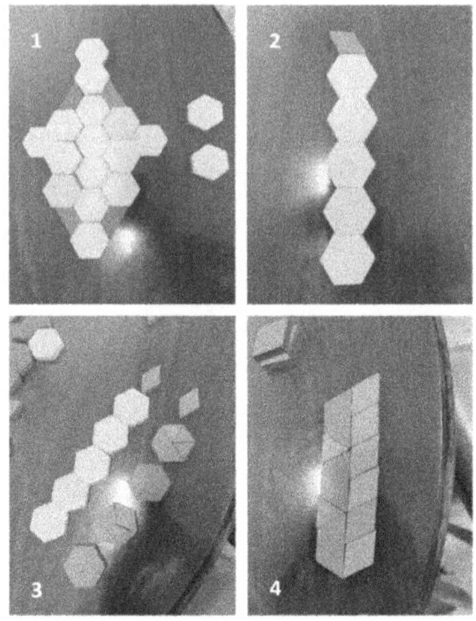

FIGURE 5.10 Pattern Block Fraction Task 2

Moving forward in the lesson, the teacher tasks the same student pairs with a more challenging problem. They are to build a pattern that represents one-third of their original pattern (see Figure 5.10).

Note that the teacher allows the use of additional blocks (blue and green) to open avenues for the creation of the fractional portion. Again, the teacher explicitly attends to an aesthetic dimension of the task. Cara and Trent work toward a one third portion by replacing pairs of red trapezoids with yellow hexagons (1), then dividing those hexagons into third portions (2). From there, the students begin experimenting with possible, equivalent assemblages of blocks (3) until they land on a pattern that they find pleasing (4). Interestingly, though, Trent is skeptical that the final pattern is a correct representation of one third of their original pattern. In an earlier chapter, we examined *the believing game* (Harkness, 2008) and how we look into our students thinking to find some truth as they perceive it. In that earlier chapter, we examined how believing in our students thinking opens space for play and exploration, but fundamentally, such believing allows us a window into our students' truths. In this instance, Trent has applied some numerical reasoning (1/3 of 27 equals 9) to their pattern. While Trent's thinking elides the differences in the blocks, there is some embedded truth in his reasoning; there is something there for we teachers to believe in and build upon. Once we are able to see Trent's truth more clearly, we can imagine tasks that allow us to arrive at truths about fractions that are mathematically valid and taken as shared. All the while, as we move toward such shared truths, we can create and enjoy beauty within the moments.

Questions for Reflection

As we conclude each chapter, I offer some questions that you might take up with your colleagues or consider on your own. Each of the virtues presented in this book is a deep and complex idea that has guided human contemplation for millennia. As such, I find it useful to sit with our thoughts and discuss them with others. As with other reflective questions, the ones I pose below do not have easy answers, but my hope is that they provide you with space for introspection which is, itself, a beautiful practice aimed at seeking truths.

- *What do you find beautiful in the world around you? What sorts of things do you find aesthetically pleasing in terms of sight, sound, smell, touch, and taste? What do your students find beautiful or pleasing?*
- *How does beauty connect to truth in your mind? Do you see a connection and if so, how? What do you find beautiful about truth, and vice versa? If you don't see much of a connection, what difference or distinction do you draw between truth and beauty? What divides them?*
- *How often, in your instruction, do you feel empowered to slow things down and take time to experience the beauty of a moment or a students' thinking? Do you feel like you have the autonomy to build in connections to beauty and truth in your math teaching? If not, why not and how might you reclaim some autonomy in your teaching?*

- *Do you think of math as something that can be beautiful? If so, how? If not, why not?*
- *Do you think your broader school community (administrators, parents, and community members) would find value in connecting mathematics with beauty and truth? If so, what do you think they would find valuable? If not, why do you not think they would find value in these ideas?*

References

Akintoye, S. A. (2010). *A history of the Yoruba people*. Dakar: Amalion Publishing.

Andrews, W. S. (1917). *Magic squares and cubes*. London: Open Court Publishing.

Fu, X., & Wang, Y. (2015). Confucious on the relationship of beauty and goodness. *The Journal of Aesthetic Education, 49*(1), 68–81.

Harkness, S. S. (2008). Social constructivism and the *Believing Game*: A mathematics teacher's practice and its implications. *Educational Studies in Mathematics, 70*(3), 243–258.

Holt, J. (1982). *How children fail* (2nd ed.). New York: Delta.

Keats. J. (1897). *Poems by John Keats*. London: Chiswick Press.

Lang, Y. (1991). *Zhongguo miexueshi dagang* [Outline of the history of Chinese aesthetics]. Shanghai: People's Publishing House.

Lu, P. J., & Steinhardt, P. J. (2007). Decagonal and quasi-crystalline tilings in medieval Islamic architecture. *Science, 315*(5815), 1106–1110.

Maheux, J. F. (2016). Wabi-Sabi Mathematics. *Journal of Humanistic Mathematics*, *6*(1), 174–195.

Merriam-Webster. (n.d.). Truth. In *Merriam-Webster.com dictionary*. Retrieved March 23, 2025, from https://www.merriam-webster.com/dictionary/truth

Muller, F. M. (2001). *The Upaniṣads*. London: Routledge.

National Governors Association (NGA) Center for Best Practices & Council of Chief State School Officers (CCSSO). (2010). *Common core state standards for mathematics*. Washington, DC: Authors.

Oluwatobi, D. E. (2022). The notion of truth among the Yoruba's. *Philosophy International Journal, 5*(1),1–7.

Piaget, J. (1970). *Genetic epistemology.* New York: Columbia University Press.

Pickover, C. (2009). *The math book.* New York: Sterling.

Seshagiri Rao, K. L. (1970). On truth: A Hindu perspective. *Philosophy East and West, 20*(4), 377–382.

Stephan, M., Cobb, P., & Gravemeijer, K. (2003). Chapter 5: Coordinating social and individual analyses: Learning as participation in mathematical practices. *Journal for Research in Mathematics Education. Monograph, 12*, 67–102.

Su, F. (2020). *Mathematics for human flourishing.* New Haven, CT: Yale University Press.

Telesco, P. (2019). Simple and compound time signatures re-examined. *Journal of Music Theory Pedgaogy, 30*(11), 1–15.

von Glasersfeld, E. (1995). *Radical constructivism: A way of knowing and learning*. London: Falmer Press.

6

Justice and Community

For each chapter, I have curated a playlist from my own music library. These are songs that put me in the mindset of the topics at hand. Some may be on the nose and others may feel like a bit of a stretch. Hopefully, though, you find the music an enjoyable addition to your thinking in these areas.

Note: Spotify automatically adds selections to a given playlist. My original selections are the first 6–8 songs on a given list

https://open.spotify.com/playlist/4NSh7c9AGm7dvb3bHxyQQB?si=BiC21tvZQwivDtdOIl2tYg

A Kentucky Preamble

As a native Kentuckian, I have come to realize that a lot of folks think of my state (a commonwealth, actually) as wholly Appalachian. While the eastern third of Kentucky is, indeed, squarely in the Appalachian Mountain range, as one travels westward, the land becomes much more gently rolling. I grew up in the south-central portion of the state where the land is level enough for large-scale agriculture and my childhood was spent on a farm that was so flat that I could see for over a mile in many directions.

My dad was one of those people who was comfortable wearing a lot of hats. He was a university professor and certified public

DOI: 10.4324/9781003484783-6

accountant, but his true love was farming. As much as he loved working with college students, driving a tractor or combine back and forth across a field brought him deep satisfaction. You could say that farming, for him, was a virtuous place. That is, he realized a connection to the things that made life worth living, working the land; he found deep fulfillment in the beauty and meaning of that work. On our farm, we grew soybeans, corn, and occasionally wheat. No matter the crop, though, my dad particularly liked night farming. After supper, he would fuel a tractor, hitch up whatever implement was called for (plow, disc, etc.) and return to the field to work the land well past sundown. Some nights, he would invite me along, and some of my most special childhood memories are sitting in the cab of a tractor on his lap, falling asleep to the soft sounds of old country acts like Glenn Campbell or The Judds. Somehow, he would get me back to the house without stirring me, and it felt like he had performed a magic trick when I woke up in my bed the next morning. Farmers, I came to realize by watching my dad, develop a very special relationship with nature. They are closely attuned to the weather. They are fiercely protective of the living things they grow. They see the sun rise and sun set most every day as they work in wide open spaces.

Farmers, at least the ones I grew up with, also operate with a deep spirit of community. The work of farming from the preparation of the fields through the harvest is filled with variables that often require the efforts of many farmers working together for a common cause. An unexpected failure of a combine, for example, in mid-harvest precipitates a call to neighboring farmers for assistance in bringing in the crop. This was certainly true in my childhood community of Rich Pond, KY (Warren County) where farmers routinely helped one another with whatever needed doing. Folks were ready, at a moment's notice, to spring into action with a tractor or toolbox to assist someone, and they didn't hesitate to make a call for help if needed. In my memory, there was no shame or aspersion to making such calls, and nothing really expected by those answering the call – albeit a home-cooked dish or baked goods often appeared after such favors had been rendered. Rather, they just supported one another through thick and thin. As child, I remember spending time at

Carters General Store and Sawyer's Mill (both local hangouts for community farmers) getting hugged, listening to stories, and, perhaps, given a Coke or candy bar by these folks who looked like giants to my young eyes. They weren't family by blood, but family by geography and purpose. Among those memories of night farming I mentioned, in many cases we were headed out to the nearby fields of some other farmer to complete work that needed doing in a timely manner. It was all the same to my dad, though. Farming was farming.

Savvy readers might notice that I have talked a lot about my dad, but a bit less about my mom in recounting thus far. What about the women in the farming community? I can resoundingly affirm that the women of Rich Pond were deeply involved in this work as well (as well as most of the adolescent young folks). Only because my dad loved driving the tractors and combines so very much, my mom would often be following close behind in the pickup truck out to the fields. She had spent years as an elementary school teacher and had a knack for finding the key tasks that needed doing and then knocking those out quickly. I have vivid memories of my mom, herself a little more than 5 feet tall, out in the fields surrounded by mature, tasseled corn stalks that were far taller than she was (see Figure 6.1). Note, my dad was 6′1″ for reference and this was a particularly good harvest due to regular rainfall during the hot summer months.

Of course, in this community focused on collaborative farming activity, mathematical connections abounded. As you might suspect, rain amounts through the summer impact the

FIGURE 6.1 My Parents in the Corn

growth of plants. Further, combating insects and disease can also protract the growing period. Typically, a harvest period in Kentucky might run from mid-August to October depending on the crop and conditions. The first hard frost in Kentucky normally occurs in mid-October but can sometimes come much earlier. Depending on how the growing season went, a farmer might be facing an early fall frost with a crop still in the field. In the best of circumstances, a harvest is, essentially, a complex mathematical problem to be solved.

Before we get into that, let me provide just a bit description regarding the mechanics of harvesting. A combine (see Figure 6.2) harvests the fully matured plants.

Depending on the crop, the language one uses would vary – you might *thrash* beans or *shell* corn and each requires a different head-unit on the machine. The basic process, though, is similar. Mature plants are cut and processed such that the salable material (e.g., the soybean and the corn kernel) is retained in a large bin behind the cab. The applicable unit for grain is called a *bushel* and is equivalent to 64 U.S. pints or approximately 36.4 liters. As the farmer harvests, the bin fills with kernels and beans and, once full, is emptied into a large truck for transport to a local receptacle at the farm, or regional grain terminal (often called an elevator) for storage and eventually sale. Really skilled farmers can

FIGURE 6.2 Combine Harvester. Shutterstock, Nikolai Tsvetkov

A single combine can harvest about 120 acres/day

1 acre yield	91 bushels
Combine Capacity	200 bushels
Grain Truck Capacity	800 bushels

Information above is for agricultural activity circa 1980

FIGURE 6.3 Harvesting Statistics Circa 1980

coordinate to drive the truck alongside the combine and transfer grain as the combine continues to harvest. Regardless of method, bringing in a crop is a delicate dance of people and machinery that depends on community and, also, mathematics.

Consider the following information (see Figure 6.3). Now, imagine this information in concert with the size of the farm on which I grew up – 900 acres.

Now, imagine that, based on how your crop was maturing, you were planning to use your combine to harvest during the last week of September and first week of October. You had planned to call the closest neighbor to see if they (or maybe their teenage kid) could drive the grain truck. Already, this is a bit of math problem regarding the harvest timeline and management of equipment. However, in the days prior to the harvest, the weather forecast calls for sustained hard frost beginning October 1 that will wreak havoc on the soybean yield. So, what my dad and others would do in those moments is map out a plan. They would coordinate their equipment, considering the mathematics involved in the harvest, to bring everyone's crops in as quickly as possible. Consider the following scenario:

> If you had 900 acres to harvest in 4 days, how many combines do you need to run? How many grain trucks do you need to ensure continuous operation of the combines assuming it takes 1 hour to unload a full truck at the farm's grain bin?

Now, if you were to talk with a real farmer and not the academic son of a farmer, that person would likely tell you about

15 other variables that could be quantified and required consideration in that scenario. All this to say, farming is a deeply mathematical enterprise, but also one steeped in virtue. Wendell Berry (2000) wrote about this in his poem titled "The Man Born to Farming". This one is easy to find via a quick internet search, and I encourage you to take a moment and give it a read. Berry, a Kentucky agriculturalist, author and poet, describes the deep fulfillment that comes with such connection to nature. As my dad and his friends plotted out the mathematics of a harvest, they were rejoicing, as Berry puts it, the *"light rise again in the corn"*. They were expressing an *"unending sentence of love"* with their efforts. And all of this, they were doing in community and with a spirit of togetherness.

Driving east from Rich Pond, KY, for about two hours would land you in foothills of the Cumberland Mountains which are, themselves, part of the larger Appalachian Mountain range. For those outside the region, if you want to sound like part of this community, it's pronounced "apple-atcha" (rather than "apple-aysha"). Because of the terrain, large-scale agriculture is not really viable; however, coal and timber abound. Coal mining, in particular, has historically served as the foundation for economic activity in this region for generations and, as such, resulted in community formations not too dissimilar from those of Rich Pond. The work is very different, but my sense of the togetherness among members of mining communities strikes me as quite familiar. Also, as with farming, mathematics frequently intersected with mining life in many different ways. I should note, though, that my sense of mountain life in Kentucky is learned rather than lived. While I had extended family that lived in areas close to the foothills, my understanding of life and community in eastern Kentucky came much later – as did my understanding of justice and injustice in these places, too.

While there is much natural beauty and rich culture in this Appalachian region, there is also a legacy of anguish that is inextricable from a vocation that steadily provided for families, built towns, but was so very dangerous and claimed the lives of many. Coal is revered as a resource, as it lent a very real sense of economic wealth to the region. However, there is a darker side of coal mining too. Tyler Childers's (2011) song, *Coal,* captures this fraught nature quite well. I highly encourage you to take a few minutes and give it a listen on your preferred streaming service. In this gut-wrenching song about tribulation, Childers mentions "mining for the company" which is a key element of coal culture in Appalachia. The places where coal was plentiful were very often the hardest to reach. Roads were poor or nonexistent and the mountains themselves were isolating for small communities like Gano (Harlan County, KY) and Glomawr (Perry County, KY). To make the economic prospects of mining viable in such remote areas, coal companies often created their own small towns near the mining sites:

> As a pragmatic solution, [mining corporations] sometimes developed a company town, where an individual company owned all the buildings and businesses . . . The remoteness and lack of transportation prevented workers from leaving for other jobs or to buy from other, independent merchants. In some cases, companies paid employees with a scrip that was only good at company stores. Without external competition, housing costs and groceries in company towns could become exorbitant, and the workers built up large debts that they were required to pay off before leaving. Company towns often housed laborers in fenced-in or guarded areas, with the excuse that they were "protecting" laborers from unscrupulous travelling salesmen.
>
> (PBS, n.d.)

Not only was the activity of mining itself very dangerous, but coal companies wielded isolation to economically exploit communities of workers, and this *unjust* practice continued for generations. Hence, mentions of "the company" in contemporary Appalachian music reflect the ways in which corporations indelibly shaped the lives of people in the region including their direct economic experiences. In 1878, a coal miner from Athens County, Ohio, wrote,

> We cannot exchange [company scrip] with farmers or others. A farmer comes to my door. He has produce, just

> what I need. He sells for thirty cents. He also wants something out of the store and would willingly give me the produce and take the [scrip] on the store but the store will not receive [my company scrip] from him so he is obliged to sell his produce to the store, and I am forced to pay the store forty cents for the article [because of store markup] I could have bought for thirty cents from the farmer.
>
> (Ohio Bureau of Labor Statistics, 1878, p. 166)

Consider the following tasks (see Figure 6.4).

This is a fairly complex sequence of tasks that would require students to purchased goods at varying costs as a portion of weekly mining income accrued over one year. Perhaps more interestingly, though, is the determination of fairness in task 3. Below are responses from Kayden and Briana (see Figure 6.5).

Kayden determined that this economic arrangement is "pretty fair" in because the difference in cost is only 1% of the miners' income. Note, Kayden does not seem to account for the task parameter of purchasing two bushels per week. Conversely, Briana asserts that this practice is deeply unfair in that this cost difference accounts for more than an entire week's work. I suspect you may already be thinking about the rich discussion we could facilitate around considering cost as a share of yearly income compared to working hours required to generate this difference. Fundamentally, though, this moves the mathematical conversation beyond the calculations and into the domain of justice.

COAL MINING WAGE (1800s):	$10/Week (worked 6 days per week)
PRODUCE (Farmer Price):	$.30/bushel
PRODUCE (Company Price):	$.40/bushel

1) **If a coal miner buys a bushel of produce twice per week from the company store, how much more money does the miner spend vs. buying it from the farmer over the course of a year?**
2) **What portion of the miner's income is consumed by this control of buying and selling goods?**
3) **On a scale of 1-10 (1 least fair; 10 most fair) how would you rate this system and why?**

FIGURE 6.4 Mining Tasks

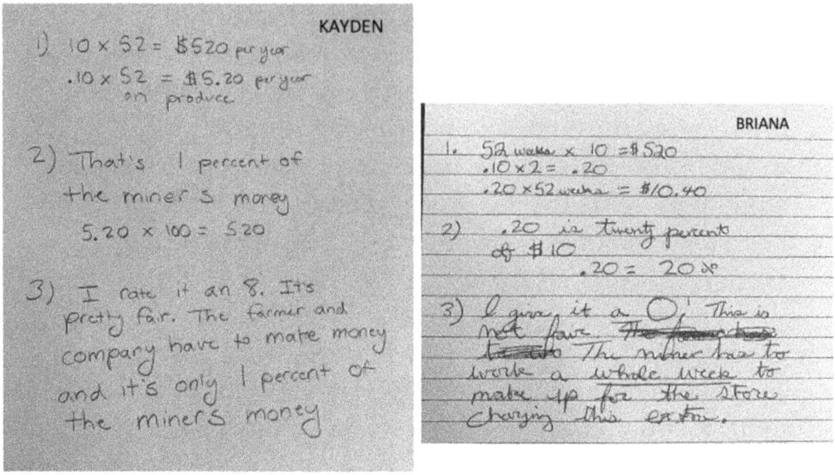

FIGURE 6.5 Mining Task Responses

Do we, as a community, consider these systems just or unjust and why? Lastly, for those who may wonder about the relevance of these sorts of experiences, Tyler Childers song, *Coal,* has over 175 million streams on Spotify. A new generation of young people (including many outside of Appalachia) are learning about the beauty and struggles of region, including influence of coal mining, through the songs of Childers, Chris Stapleton, Tony Logue, Sturgill Simpson, and many other contemporary artists that carry on the rich legacy of mountain music.

Lastly, I would be remiss if I didn't account for coal production in the western regions of Kentucky. If you were to drive about one hour *west* of my hometown of Rich Pond, KY, the land would not be nearly as mountainous as the eastern portion of the state, but you would be in coal country, nevertheless. In his song, *Paradise*, legendary singer songwriter John Prine (1971) wrote of coal mining in western Kentucky. Again, I encourage you to take a few minutes and listen to Prine's beautiful recollection of childhood experiences "down by the Green river" and how those spaces and places have been "hauled away". In addition to the historic economic injustices, the company towns, Prine sang about environmental impact of coal on his homeland and how, once the coal extracted, land and communities were forsaken. How folks were left behind in diminished lands with diminished prospects. "Mr. Peabody's coal train" is a reference to the Peabody Energy company which owned and operated numerous mines in the region and quite literally hauled away wealth and beauty from the area. Prine's lyrics provide additional layers to discussions of justice as it relates to coal mining and the communities that mining shapes. I encourage you to step away for a few minutes, search for these songs, and give them a listen. To hear from the strong women of the mountains, perhaps add *Coal Miner's Daughter* by Loretta Lynn, *The Way I Talk,* by S.G. Goodman, or *You'll Never Leave Harlan Alive* by Patty Loveless to your playlist. Sit a spell with these songs. Imagine the people and the communities behind them. Let the beauty, history, richness, and struggle of my beloved Kentucky embrace you:

What songs make you feel most connected to your home places, your cultures, and your histories? What songs speak to struggles in those places? What songs would your students name? Where might mathematical ideas intersect with these songs? Jot down your notes here.

____

____

____

____

In the sections that follow, we will examine the virtues of justice and community as they connect to mathematics. We will explore how these virtues might inform our teaching in the elementary mathematics classroom and what practices we might deploy to bring these ideas to light. Certainly, the mathematics of agriculture or coal mining may not reflect the particular lived experiences of your own students. While we can always share experiences beyond those which we have lived, we may also search within the cultures, histories, and communities of our own classrooms to find mathematical opportunities. Indeed, teaching mathematics in a way that precipitates justice and community is not only possible, but desirable in that such teaching enriches the learning experience by connecting it to deep human pursuits.

The Virtue of Justice

As with other virtues we have explored in this book, justice is a human pursuit that appears universal across time and geography. Depending upon the thinker, ancient philosophies of justice may appear quite crude. For example, Polemarchus, an Athenian scholar who lived in the 5th century BC, wrote of justice as "doing good to friends and harm to enemies" (Lycos, 1987, p. 21). However, Socrates and his protégé, Plato, had more nuanced and expansive visions of justice as a key element of social harmony and human flourishing. From a Socratic perspective, "justice is something personal, an internal harmonious ordering of the human soul . . . and not an external social convention to keep some order and peace in the everyday dealings" (Evangeliou, 2007, p. 67). Further, these more enlightened Greeks considered justice as an active pursuit of the individual seeking who strives for goodness by giving to society and receiving from society proportionate to what is given. Moreover, it is in the giving of oneself toward some greater societal good that generates conditions for fulfillment. John Rawls, a contemporary legal philosopher and theorist, referred to justice as the "first virtue of social institutions" and considered the principle of *fairness* as essential to its nature (1999, p. 3). Rawls goes on to write,

> Many different kinds of things are said to be just and unjust: not only laws, institutions, and social systems, but also particular actions of many kinds, including decisions, judgements, and imputations. We also call the attitudes and dispositions of persons, and persons themselves, just and unjust.
>
> (p. 6)

Ultimately, though Rawls describes "justice as fairness" regarding the interactions among individuals as well as societal principles, it examines how such justice intersects with notions of equality (p. 11). In keeping with Plato, he describes justice as something that is both produced and received by individuals. In other words, we all have a hand in creating justice in the world around us, and we all benefit from the just actions we take and systems we make.

Of course, fairness, as a concept, is often quite subjective. When my children were younger, my daughter would dispute the fairness of her bedtime with respect to her brother to whom she was three years younger. She would argue vehemently that fairness should be expressed in equality of bedtime such that she would be allowed to stay up an hour longer to match the bedtime of her brother – their treatment should be equal which would precipitate bedtimes that are the same. This common conception of equality (sometimes called ordinary equality or descriptive equality) refers to a fundamental sameness. We can consider two different words equivalent if they "have both a single and constant meaning in all usages" (Westen, 1990, p. 9). However, this is trickier than one might suspect. Consider the different ways we might use the word *square* – square meal, square deal, square dance, and of course, to describe a particular type of quadrilateral. Similarly, *bedtime,* may not be equivalent as it retains varying contextual features specific to each of my children. When we, as individuals, put equality into practice (in pursuit of justice as fairness), the complexities of the human experience complicate our efforts immensely. Imagine you want to prepare snacks for a math small group that is comprised of five children who did well on their assessment. Unfortunately, you didn't have time to go to the grocery, but you have enough food on hand to make a tasty snack to treat each student. Sounds simple, right? However, when it comes to fairness, things quickly become complicated. You have recently become aware that one your students' family has recently become food insecure. His performance on the assessment was the lowest in the group. Another student did extremely well on the assessment but bickered with others every day to the point of disruption. Another student had the second-lowest score on the assessment but was highly engaged and asked great questions. Likely, you see where I am going with this. In this scenario, you might still elect to provide each child with a single snack, but is this the fairest outcome? Does this outcome produce the greatest justice? If you are like me, then you probably answer no to these questions. How, then, do we distribute the snacks though? If not equally, what is the most just distribution? Even if we do decide to give each child a single snack, our reasoning for this decision is what legal scholars refer to as *prescriptive equality* (Westen, 1990); This is, in other words, equality that is attached to some human judgment. While equality and fairness are often used somewhat interchangeably, they are frequently quite different when it comes to the manifestation of justice. Indeed, equality can sometimes masquerade as justice while being highly unjust in practice – fair isn't always equal, and equal isn't always fair.

So how do we, as educators, create classrooms that reflect an ethos of justice? Toward this end, Su (2020) writes, "justice means treating people equitably – giving each person what they are due. It means speaking up for the powerless, who are most likely to be treated unjustly" (p. 149). Justice in our instructional spaces is an active pursuit and one which often requires judicious decision making and advocacy outside of our comfort zones. Su continues,

> Some speak of justice in two categories: *primary justice* and *rectifying justice*. Both are important. Primary justice involves right relationships: treating each person with dignity and care and establishing social practices and institutions that support those aspirations. We flourish when we treat each other justly and when we are treated justly. Rectifying justice is spotting something wrong and trying to make it right. If primary justice were commonplace, rectifying justice would not be needed. But injustices are everywhere. Coercive power relationships are destructive. People and institutions may subtly promote inequities without realizing it.
>
> (p. 150)

What Su is describing is a discipline of practice that implores us to not only to strive for the most compassionate and empathetic stance toward others, but also to seek places where we might better the lived experiences of those unjustly treated – "spotting something wrong and trying to make it right". But how do we know what's wrong and right? Jesuit Catholics would refer to this process as *discernment*. Abigail Gosselin, a philosophy scholar, writes, "discernment, broadly, speaking is a process aimed at making good judgements" and this involves attending closely to one's thoughts, feelings, and intuitions as they relate to decision making (2012, p. 16). In his *Letter from a Birmingham Jail,* Dr. Martin Luther King Jr. (1963) described a process of self-purification aimed at preparing members of the civil rights movement for non-violent, direct action to address segregation laws specifically and racial inequality more broadly. Deeply embedded in this process of building personal resilience to absorb injustices and violence during protest activities was a continuous discernment of justice and how such justice may separate from policy and legality. King writes, "Any law that uplifts human personality is just. Any law that degrades human personality is unjust. All segregation statutes are unjust because segregation distorts the soul and damages the personality". Reflecting on these ideas in terms of our teaching, our task is to engage in meaningful discernment such that we may find the most just and humane ways of interacting with our students while also seeking to address injustices wherever they may occur in our instructional spaces as best we can.

The Virtue of Community

Attuning ourselves to enacting primary justice and seeking pathways toward rectifying justice naturally lends itself to the formation of community. Community, in its simplest form, is a group of people bound by relationships or something shared. Treating students with dignity and respect tends to engender trust and relationships. Working toward bettering the unjust experiences of those in our classrooms builds similar bonds. Moreover, rectifying justice doesn't live only within the walls of our classrooms. We may experience community in a shared purpose of learning how math can be deployed to not only understand injustice but also actively make the world a better place. At the center of these processes and pursuits is the formation and strengthening of a classroom community where students within feel safety and empowerment such that they may practice discernment and bring mathematics to bear on their lived experiences and society more broadly.

I imagine that the importance of community in the math classroom seems somewhat obvious to many readers. It's very challenging to teach well without developing relationships and shared goals with our students. However, like other virtues we have explored, community is something that great thinkers across time and place have identified as a key element of human flourishing. Confucius, in approximately 500 BCE, developed a philosophy of *Ren* which means to love our fellow humans and to seek ways to be helpful to those around us (Shi, 2008). Around that same time, Aristotle likened community to the human body "while the individuals are the parts of the body. The whole body may function but not to its fullest if one or two parts are not in order, are removed or destroyed" (Onwuatuegwu, 2020, p. 62). The importance and value of individuals coming together to share goals and experiences permeates most every cultural, philosophical, and religious tradition. For example, for tens of thousands of years, Indigenous Australian peoples organized to form communities, not only to enhance prospects for survivability, but also the human experience of sharing language and culture with others (Dudgeon et al., 2002). From time immemorial, human beings have realized the deep importance and fulfillment that comes from building relationships and the sharing of ourselves with others.

Unfortunately, though, communities may be inaccessible to some, and this is particularly true, historically, of mathematical communities. Su (2020) writes,

> mathematical communities can be very hierarchical, even if they don't wish to be . . . in a classroom setting, learning math can be a challenge in community because the skill is at the center of instruction. So, if group work

> isn't designed well, students who need time to think will be dismayed by those who finish first. The singular focus on skill, even when it's warranted, leads people to idolize those perceived to be stronger in that skill, and creates unnecessary hierarchies in any mathematical community.
> (pp. 191–192)

What Su describes is the way that mathematics instruction is often experienced as divisive and disruptive to the formation of community, given the centralized focus on skill. Often, in elementary classrooms, this tends to manifest as the emphasis on correct procedures and answers at the expense of those who find value in process and thinking. Su continues, "That feeling – *I don't belong* – can be quite crippling. And this is where community is terribly important: for us to feel belonging" (p. 188). As teachers of mathematics, one of our first goals should be to ensure that we create classroom cultures where our students feel they belong. Without some sense of belonging to a community, even the best instructional practices will not be able to reach those who feel that they are on the outside looking in.

Building a Culture of Justice and Community

With the importance of justice and community in mind, the challenge for us as educators is to design a mathematics classroom that engages with these virtues. How, though, do we do this when our students have different lived experiences? Our students may not all share a native language. Cultural customs and traditions may vary. How do we create something we all share among individuals who, at first blush, seem so different? To this, I would argue that we, as human beings, all share a desire to be a part of something. There is a reason Confucius, Aristotle, and Aboriginal Australians concerned themselves with community all those years ago – because it is fulfilling. As mathematics teachers, we are in a unique position because we have something on which to focus. As we have explored in other chapters, mathematics can create beautiful, joyful places for play. Math can help us find freedom and meaning in the world around us. Given the power of mathematics to create space for other virtues, it can also provide a foundation for the formation of community and the enactment of justice. I don't mean that math is some universal language that transcends culture. Rather, I intend something quite different – that math is a deeply human experience and one that we can share with our students as a place for togetherness, respect, dignity, and improving the world around us.

From this perspective, we can rely on other virtuous mathematical connections to develop our classrooms in this direction. No matter how diverse our classrooms may be, likely we can agree that finding meaning in a moment is important. In our examination of meaning and perseverance, I referenced the concept of disequilibrium (Carter, 2008). In that chapter, I shared how such healthy struggle is essential to the construction of mathematical meaning; however, building a culture around the honoring of process and struggle is also a very real exercise in community building. Carter writes,

> whenever the class worked on a new or difficult concept as a whole group, I reminded students of the value of disequilibrium in our mathematics class. Students assessed their own progress by giving a thumbs up if they were thinking, 'I get it and I am feeling very comfortable', thumbs sideways if they were thinking, 'I have disequilibrium and I am struggling'; or thumbs down to signify, 'I am totally lost, and have no idea what to do next'.
>
> (p. 136)

More than just providing an actionable way to informally assess mathematical struggle in the moment, creating space for students to share their level of disequilibrium allows for the centering of meaning-making rather than answers. Moreover, when we stop to lift up healthy struggle as necessary for learning, we create a classroom where all can be a part of the mathematical journey as we focus on adding and removing scaffolds to ensure each student has an opportunity to grapple with concepts. Carter continues, "I stopped the lesson to discuss disequilibrium. 'This is the feeling of disequilibrium', I explained to my first graders, 'and now the real learning can begin.' Our motto became, 'if you are not struggling, you are not learning'" (p. 136). Here, Carter illustrates how the virtue of perseverance provides platform on which we can build community. We can all share in the rewarding process of healthy struggle, including teachers. Su (2020) describes how we, as teachers, can also share our moments of mathematical struggle in the form of mathematical autobiographies or *mathographies*. For example, "teachers who ask students to write their "mathographies" – their autobiographical journeys in math experiences – often begin by sharing their own. Leaders who are vulnerable can help other overcome their feelings of being an imposter" (p. 195). This sharing of ourselves can erode the hierarchies that often occur in mathematical spaces as we explore the differences and commonalities of our stories. For younger students, these mathographies might take the form of imagery (e.g., draw a picture of how math makes you feel and be ready to tell me about it). No matter the format, the sharing of our experiences and feelings, teachers included, is one way to find togetherness and community in the math classroom. Lastly, building a culture of community is not limited to our students. The more we can involve other stakeholders (e.g., families and administrators) in our math experiences, the richer and more supportive the texture of those experiences becomes. As teachers, we should maintain a posture of welcome and outreach to all who connect to our classroom communities and endeavor to extend our communities outward. As a parent, I know that families are so often inundated with newsletters, daily emails, and other school communications that can blur into noise due to the sheer volume. As pedagogical engineers, though, we can design around this. Rather than a long email out to families, perhaps we send a single prompt such as, "Pick a number, any number, and sing a silly song about it with your student", "What is your favorite math word and why?", "Do you feel like you belong in math class?" These prompts don't necessarily advance a particular content standard or skill but rather allow important activities and conversations (even silly ones) to extend beyond the classroom. More importantly, they signal to our stakeholders a desire to involve them in our communities.

In the same way that meaning can provide a foundation for community, power can provide a foundation for justice. First and foremost, returning to the notion of primary justice, we may consider how power is apportioned among students and teacher. Are students properly empowered such that they are afforded dignity and care? Are we actively looking to create opportunities for our more hesitant students to enter into a discussion? Are we shifting power away from students who, sometimes, have the tendency to dominate the discussion and talk over their peers? Being mindful of how power flows throughout the classroom is a core mechanism for ensuring primary justice where individuals within the community feel respected, honored, and welcome.

Rectifying justice is an area where we can connect principles (e.g., respect for persons and equitable access to resources) and mathematics to better the world around us. Sometimes this form of justice can drift into platitude. Sure, let's try to fight injustice wherever it occurs. Now what's the next thing in our lesson? Rather, let's name a few places, specifically, where we, as a community, might place our attention. Beyond the classrooms, there are plenty of injustices that might provide terrain for math activity – wealth inequality due to structural racial discrimination in zoning (e.g., redlining), disparate legal outcomes and rates of incarceration among varying racial groups, and so forth. However, let's bring this conversation a little closer to home and consider how we might restoratively make our classrooms more just. Gutierrez (2009), provides us with four lenses through which we might look for these opportunities: access, achievement, identity, and power.

An honest appraisal of access in my classroom might reveal that my students who are non-native speakers of English consistently do not have adequate access to the curriculum because of my reliance on English text to generate context problems. On the achievement front, I might find that the several Ukrainian students who recently arrived in my classroom consistently score poorly on summative assessments despite trying very hard to engage with my lessons and often doing well on classroom assignments. Regarding identity, perhaps I see in some of my students' mathographies that they consistently feel like they don't belong in a math community (or even in the school community, for that matter). Lastly, in terms of power, I find that I too often let the same few students dominate the discussion just because they are so eager to contribute. Each of these examples contains an opportunity to enact some rectifying justice. As the teacher, I can take action to make the math experience more equitable by opening up pathways for access, revising my assessments to more accurately reflect mathematical capability, by creating opportunities for creativity, play, and success such that marginalized students may feel more like mathematicians, and lastly, to distribute power more evenly and lend voice to those who have been sidelined.

I can also zoom outward in terms of restorative justice. At my core, I am an optimist and have great affection for my commonwealth of Kentucky and my home country of America. However, to deny the persistence of systemic injustice is to deny history, current reality, and the ways the two intersect. Poor water quality and the prevalence of food deserts (i.e., lack of access to affordable, nutritious food) in both urban cores and Appalachian counties are injustices that we can use math to explore. Imagine a map of your town or city divided by zip code. If you were to live anywhere in that zip code, how many grocery stores would be within 1 mile of the given address? How would you determine the average distance to a grocery store from any given address? How would you rate the importance of this difference? Below are some student responses to these questions (see Figure 6.6). Note the zip codes in this task were color coded, hence the reference to green and red.

What do you notice about these responses? What do you wonder? Both students identified the green zip code as more populated with stores; however, Briana and Kayden each took a different route to determine the average distance to a store. Which method do you find more compelling and why? They also arrived at different places regarding the importance of this issue.

Fundamentally, though, tasks such as these create opportunities to turn our mathematical gaze toward justice beyond

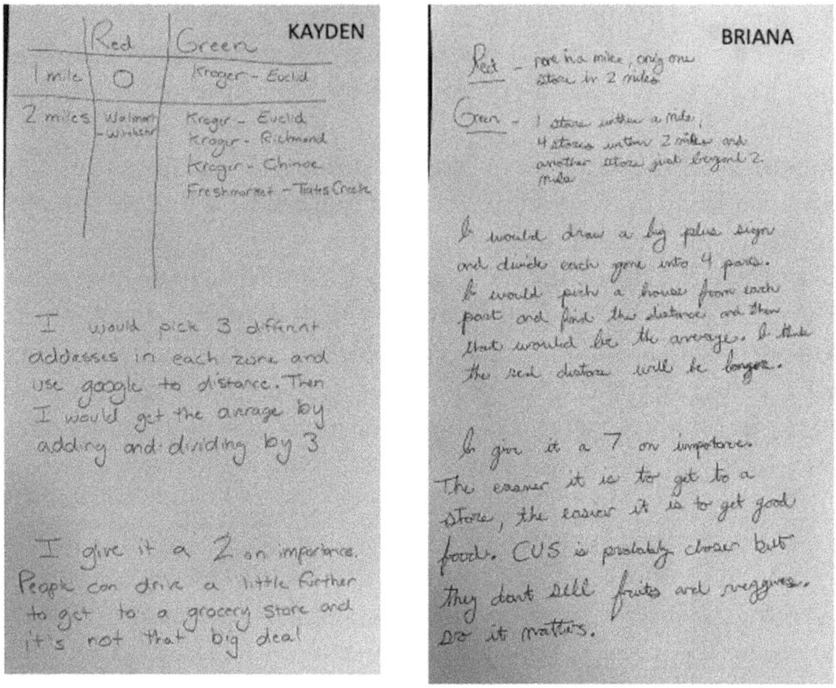

FIGURE 6.6 Student Responses to Grocery Tasks

our classrooms, to consider how we feel about these challenges, which, in turn, opens places for discussion about why such injustices (food deserts, in this case) persist. We can also use these contexts to examine how legacies of racial injustice, including redlining practices, fueled such inequities.

For those who may be unfamiliar with the term, redlining, this refers to the racist practice of denying housing loans and other financial services to minority groups within specific "redlined" areas. The consequence of this was to create and perpetuate systems of segregation and economic oppression, as wealth generation via the acquisition of housing was unavailable to black folks and other non-white groups within these redlined areas. As one might suspect, historically redlined areas in towns and cities often remain economically disadvantaged in many ways including access to healthy foods. In the same ways that early coal economics solidified and perpetuated oppression, so did the racist redlining practices within towns and cities. Lastly, I would note that redlining was formally banned by the Fair Housing Act of 1968; however, the practices persisted informally through the late 1970s (Federal Reserve History, 2023). To younger readers, this might sound like a long time ago, but as someone who was a school-age child at the time, I think of the parents of friends from school and the direct, structural racism they likely experienced as they tried to buy a home in my community.

More optimistically, we can also use these contexts to engage our students in solutions and what might be done to help. In my community of Lexington, our city government, in partnership with Kroger (a grocery chain) and our local food bank, recently began operating several large trailers, called *Mobile Markets,* to distribute (free of charge) fresh produce, meat, dairy, and other nutritious foods throughout the city, with a particular focus on districts that have poor food access. This generates new horizons for math reasoning. Where should the trailers stop and why? How much food is needed to stock each trailer and how should the food types be apportioned? What math-based recommendations could we make to our city leaders that might make the Mobile Markets more impactful? This is but one example of how a new city resource allows my students and I to combine restorative justice and mathematics in a powerful and optimistic way.

Core Practice: Painting the Picture Bigger

As I mentioned, there are many spaces within our classrooms and schools where we may practice (or increase) primary justice and find opportunities for restorative justice. A lot of these spaces may fall within our purview as leaders of those classrooms. As I mentioned, my questioning patterns or assessment practices, for example, can have real impacts on the level of just experiences my students have. We can also turn our students gaze toward unjust systems, practices, or histories in our schools, districts or communities as Kayden and Briana did with their examination of food deserts. I understand that using the math classroom as a space for social critique can be challenging on several fronts. It can feel disconcerting for administrators, parents, and other community members to hear that students, and particularly young students, are using math (something these individuals may perceive as value-neutral) to find flaws and negative outcomes in the world around them. I have found that there is a deep societal desire to ensure that our students, again, particularly the younger ones, are exposed to positive experiences and shown the goodness of the world around them. As a parent, I deeply relate to this sentiment; however, I would argue that using math for any manner of social critique (to advance primary or restorative justice) is one of those good things. The key, here, is to ensure that there is an opportunity to carry the process toward identifying solutions or ways to make situations better. Obviously, we can't go back to the late 1800s and envision rectifying justice for coal camp economics, but we can trace those histories forward to lingering economic oppression and lack of resource availability. Poor water quality is but one lingering reminder of such oppression and environmental disregard in Appalachia to this day. Returning to food deserts of the present, in this sense, the next step for Briana, who described a lack of grocery access as unfair, to vision ways to make this reality more equitable, and to use math to add grounding to such visions. Indeed, Briana may have some really good, mathematically-based ideas for making the Lexington Mobil Markets even more effective. By my estimation, this is one of the most positive and hopeful things we can do as teachers – giving students tools and opportunities to use their minds to imagine a better world.

There are so many ways we can engage in this rich and rewarding work. Perhaps you might be familiar with the children's book, *If the World Were a Village* (Smith, 2017). If you aren't, take a moment and conduct a quick Google search. In a nutshell, the book distills many different global statistics (e.g., language, educational attainment, and access to electricity and potable water) down to a village comprised of 100 people. Smith writes,

> The village stirs and comes to life, ready for a new day. Who are the people of the global village? Where do they come from? Of the 100 people in the global village, 60 are from Asia, 16 are from Africa, 10 are from Europe, 8 are from

South America, Central America (including Mexico) and the Caribbean, 5 are from Canada and the United States, 1 is from Oceania (an area that includes Australia, New Zealand, and the islands of the south, west and central Pacific.

(p. 8)

Haruto and Leisha were tasked with constructing a representation of these data (see Figure 6.7). This task is consistent with first grade mathematics content focused on organizing and representing data with multiple categories (NGA & CCSSO, 2010 – see standard 1.MD.C.4).

As you consider these representations, what do you notice? What do you wonder? Which student's representation better conveys the data? Which is more engaging? What might you like to ask Haruto and Leisha?

Moving further into the book, we might reconnect with our discussion about access to food, only this time at a global scale. Smith (2017) writes,

> All in all, there is no shortage of food [in the global village]. If all the food were divided equally, everyone would have enough to eat. But the food isn't divided equally. So, although there is enough to feed the villagers, not everyone is well fed. 33 people in the village do not have a reliable source of food and are hungry some or all of the time. 11 people are severely undernourished and are always hungry.
>
> (p. 17)

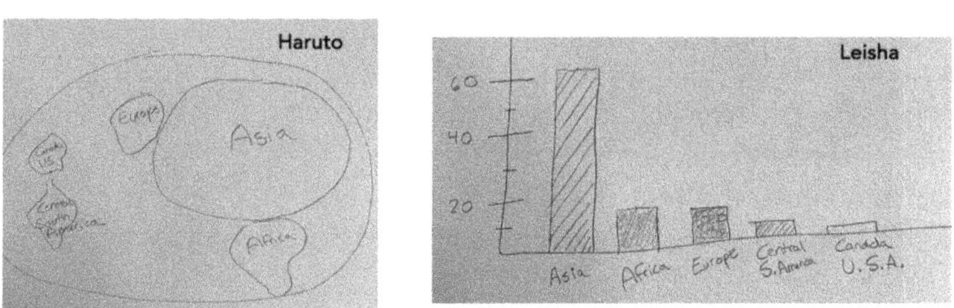

FIGURE 6.7 Student Representations of Population Data

While these statistics provide an opportunity to represent specific data, they also occasion students to identify food production sources, geographically, where surpluses occur, and use mathematics to imagine a more equitably reapportioned food supply. What might a "Global Mobile Market" look like? How many trailers would we need to address food insecurity for the most desperate country? How far would they need to travel, and so forth? While the statistics, by this point, are likely somewhat dated, the central conceit of a global village converges the notion of community and opportunities to explore justice within this broad community of the world. It's one way to paint the picture bigger, expand our field of vision beyond our math classrooms, and think about our place on the planet and our responsibilities in society, where we identify injustice, and, most importantly, how math might be harnessed to envision a better reality.

Professional Noticing for Justice and Community

Returning, yet again, to our connecting practice of professional noticing, remember that this is comprised of three components. First, we attend to the features of a mathematical moment, then we interpret that moment to find meaning and ascertain students' understanding of the math at hand. Lastly, we make some decision that advances us toward our math goals or helps us better know how students are thinking. Before we get into professional noticing, though, let's set the stage for some mathematical thinking in the context of justice and community.

In addition to economic oppression occasioned by industrial exploitation of geographic isolation, coal companies also sought to ensure that members of the mining community did not organize and advocate for more just treatment. Dr. Althea Webb, an Associate Professor of Education Studies at Berea College has written about the way these companies hired and assigned their workers around the turn of the 20th century. Webb (2013) writes,

> The African American coal miners in Central Appalachia encountered better conditions than those in the South

> and were not excluded to the same degree as those in the North. The coal fields of Central Appalachia were a major destination of African Americans leaving the Deep South during the Great Migration, and between 1870 and 1930 the African American population in Central Appalachia increased dramatically. The coal mining industry actively recruited African Americans to work alongside native white Appalachians and immigrant workers from Europe. The motive behind having three relatively equal numbers of men in the different groups of workers was to prevent unionization, as there would be natural language barriers

Perhaps you are surprised, as I was a while back, to learn that the mining workforce of Appalachia in those times was deliberately diversified by company leaders, albeit for awful reasons. Not only were the communicative channels of these miners constrained for labor management reasons, it also imperiled their day-to-day work, a good portion of which was mathematical in nature (e.g., blast positioning and cart-loading for ease of movement). While our students will never experience that particular variant of organized oppression, we can use a mathematical task to underscore the importance of communication in mathematical reasoning within a small community or group (Figure 6.8).

In this instance, the teacher has shared some of this history regarding past Appalachian mining and then presents the following task. Consider this task and the exchange around it (see Figures 6.9 and 6.10).

Again, this task is not intended to simulate the deep injustices inflicted about racial groups in Appalachia. *Moreover, I would caution teachers to be very, very careful, and to probably avoid, any attempts to simulate or replicate injustice and oppression in the classroom.* Indeed, that strikes me as a highly counterproductive and harmful activity with respect to the justice we aim to further and community, we aim to create; moreover, it serves to minimize and even mock those who experienced such injustice firsthand. Rather, this mathematical experience is aimed at shining a light on how communicative channels (or limits of such channels) impact our ability to think with one another in community. Key

FIGURE 6.8 Miners of Lorado, WV
Photo Courtesy of U.S. Library of Congress: https://www.loc.gov/item/95509007/

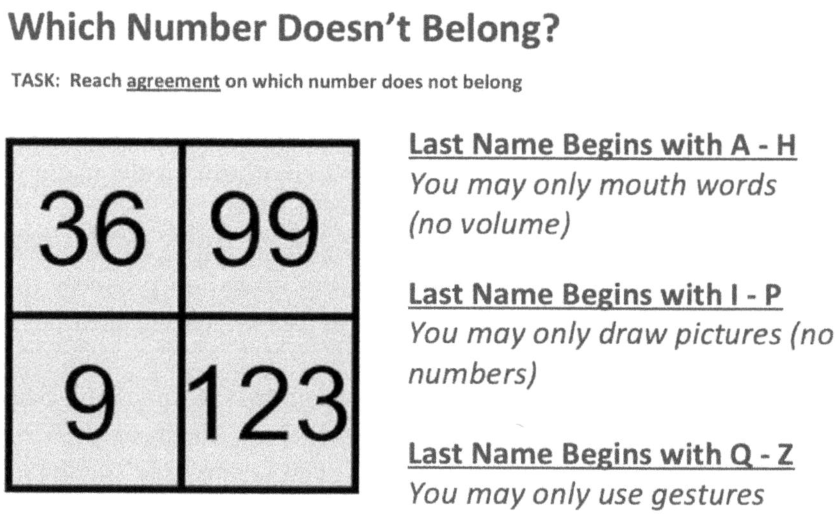

FIGURE 6.9 Limited Communication Task

to this moment is some discussion about why communication in mathematical spaces is so important. I should not intend to mimic the deep oppression and injustice of coal miners (if that were even possible), and as a teacher, I need to be cautious, clear, and explicit when wading in these waters.

TEACHER: *"So this seemed like it was tricky. I wanted you all to try and come to an agreement on which number doesn't belong, but you couldn't just talk about it. Remember, this is what Appalachian miners had to do daily and find ways to communicate about things that involved math, but didn't all share a way to do that. You all had it easier because their lives depended on it, but even here in our math time, I saw that this was still hard for some of you all. So, talk me through your process.*

SEEMA: *"I had to draw and that's really hard to draw why something is, I mean should be the one. Like, I wrote the 9, but then I had to try and draw an explanation of a single digit"*

TEACHER: *"Oh, so the 9 doesn't belong because it's only a single digit. Somebody else from Seema's group, did you agree?"*

LIN: *"Sort of, with the digits, but I was thinking the 123 didn't belong. I had to gesture, and I was just doing three, three, three with my fingers, but I don't think my group got what I was doing"*

THOMAS: *"Same, I was 123, but mouthing "it's three digits and the others are smaller. Also, I can kinda relate the other three numbers to nine."*

TEACHER: *"Tell me more about that."*

THOMAS: *"Two of them have nines in them and 36, if you add the digits, makes a nine. Also, there's factors, nine times one, times four, times eleven. 123 doesn't really relate to nine, so it doesn't fit. And it's three digits."*

TEACHER: *"So it sounds like you all didn't quite agree, but had some common focus on digits. What kept you all from coming to some agreement you suppose?"*

SEEMA: *"We just had trouble talking. I mean, not talking but, like, communicating. We couldn't talk about why."*

TEACHER: *"You say communicating, why is that important to people?"*

SEEMA: *"I mean, it's just math, but if you can't talk, you can't get stuff decided. You can't do anything. People need to be able to talk to each other."*

FIGURE 6.10 Student Exchange around Limited Communication Task

With that guidance in mind, turning to the exchange, first, we might attend to the students' arguments regarding the number that doesn't belong. Seema focuses on the 9 because it is the only single-digit number. Thomas, however, identifies 123 as the number that doesn't belong and mounts an argument about relating to 9s, including 9, 36, and 99 as being multiples of 9. The teacher, though, seems interested in other elements of the moment, specifically those involving the communicative process and probes Seema's descriptions of challenge in finding agreement. That is, the teachers' interpretation of this moment is likely organized around a perceived opportunity to engage this group in conversation about how communication can enhance the mathematical moment as well as relate to historic experiences of Appalachian miners. We see this in the teachers' decisions to probe Seema's language. Ultimately, Seema responds that such communicative constraints hinder our ability to make decisions – in math class, like most everywhere else, "people need to be able to talk to each other". This assertion provides a most excellent springboard to talk about language and communication in our own classrooms. How do we make sure that we, as a community, create paths for everyone to communicate with one another, particularly if we have community members for whom English might not be their native language? What action can we take to realize a more inclusive experience, to create a more just environment where everyone's ideas can be shared and considered? For some, this direction of professional noticing might seem to drift away from the mathematics at hand; however, if we consider doing math to be a humanized pursuit of virtue, the extent to which we are able to interact with one another as we explore our thinking is elemental to our work in the math classroom. Sharing effectively with one another is the backbone of our math community.

Questions for Reflection

As with the preceding chapters, I conclude our exploration of justice and community in the math classroom with some questions that you might take up with your colleagues or consider on your own. Issues of justice and community in schools can be complex places where you may find competing perspectives and voices. I encourage you to consider these questions, share them with a friend or colleague and talk about your responses. As with other reflective questions, the ones I pose below do not have easy answers, but my goal is that they help you examine your own thoughts about these ideas of justice and, perhaps, engage in a bit of community as you do so.

- *Where do you perceive injustice in your school, school system, or broader community? Would your colleagues agree? What is needed for restorative justice in these spaces? What role might you play? Where might mathematics intersect with such restorative justice?*

- *How would you describe the community of your math class? What adjectives would you use? Would you describe it in positive terms? If so, what do you think makes your math classroom a positive space for community? If not so positive, what do you think is missing from your math classroom in terms of community?*
- *How might your broader school community (including parents and other stakeholders) feel about the ideas in the chapter? Do you think they would see a natural connection between justice, community and teaching math? If not, how might you help these groups better understand these ideas? What are the challenges and hazards in doing so?*
- *How often do you feel that your math teaching intersects with justice? If quite often, what drives you to direct your teaching in that way? If not very often or not at all, why don't justice and math teaching connect in your professional practice? Do you feel that the ideas in this chapter are attainable to you?*
- *What do your students find unjust or unfair in their experiences? What makes your students feel a part of a community? Have you ever asked them these questions? If not, what do you think they might share with you?*

References

Berry, W. (2000). The man born to farming. *World Watch, 13*(5), 33.

Carter, S. (2008). Disequilibrium and questioning in the primary classroom: Establishing routines that help students learn. *Teaching Children Mathematics, 15*(3), 134–137.

Childres, T. (2011). Coal [Song]. *On Bottles and Bibles.* Self-Released.

Dudgeon, P., Mallard, J., Oxenham, D., & Fielder, J. I. (2002). Contemporary Aboriginal perceptions of community. In A. Fisher, C. Sonn, & B. Bishop (Eds.), *Psychological sense of community: Research, applications, and implications* (pp. 247–267). Boston, MA: Springer US.

Evangeliou, C. (2007). Socrates conception of justice: Beyond Aristotle and Rawls. *Phronimon, 8*(2), 59–71.

Federal Reserve History. (2023). Redlining. In *Federal Reserve History.* Retrieved July 14, 2025, from https://www.federalreservehistory.org/essays/redlining

Gosselin, A. (2012). Cultivating discernment. *Jesuit Higher Education: A Journal*, *1*(1), 16–30.

Gutierrez. R. (2009). Framing equity: Helping students "play the game" and "change the game". *Teaching for Excellence and Equity in Mathematics*, *1*, 5–6.

King, M. L., Jr. (1963, April 16). Letter from Birmingham Jail. In *The Martin Luther King, Jr. Research and Education Institute*. Stanford University. Retrieved July 8, 2025, from https://kinginstitute.stanford.edu/letter-birmingham-jail

Lycos, K. (1987). *Plato on justice and power: Reading book I of Plato's republic*. SUNY Press.

National Governors Association (NGA) Center for Best Practices & Council of Chief State School Officers (CCSSO). (2010). *Common core state standards for mathematics*. Washington, DC: Authors.

Ohio Bureau of Labor Statistics. (1878). *First annual report* (pp. 156–192). Columbus, OH: Nevins & Meyers, State Printers.

Onwuatuegwu, I. N. (2020). Concept of community: Aristotle's perspective. *International Journal of Research and Scientific Innovation*, *7*(7), 60–64.

PBS. (n.d.). *Company towns*. Slavery by another name. Retrieved July 8, 2025, from https://www.pbs.org/tpt/slavery-by-another-name/themes/company-towns/

Prine, J. (1971). Paradise [Song]. *John Prine*. Atlantic Records.

Rawls, J. (1999). *A theory of justice: Revised edition*. Boston, MA: Harvard University Press.

Shi, Z. (2008). Confucius's concept of Ren and its application in education. *English E-Journal of the Philosophy of Education*, *3*(1), 1–12.

Smith, D. J. (2017). *If the world were a village* (2nd ed.). Toronto: Kids Can Press.

Su, F. (2020). Mathematics for human flourishing. New Haven, CT: Yale University Press.

Webb, A. (2013). African Americans in Appalachia. *Oxford University Press* [Electronic Publication]. Retrieved from https://oxfordaasc.com/browse?t0=AASC_Subjects:68

Westen, P. (1990). *Speaking of equality: An analysis of the rhetorical force of equality in moral and legal discourse*. Princeton, NJ: Princeton University Press.

7

Love and Permanence

For each chapter, I have curated a playlist from my own music library. These are songs that put me in the mindset of the topics at hand. Some may be on the nose and others may feel like a bit of a stretch. Hopefully, though, you find the music an enjoyable addition to your thinking in these areas.

Note: Spotify automatically adds selections to a given playlist. My original selections are the first 6–8 songs on a given list

https://open.spotify.com/playlist/4o1PqTrqnQS98uaCHinh73?si=jVA9pIB-T3CN1Rgs4s2Alg

Introduction

There are lots of ways to give and experience love. We can love the individuals and animals in our families. We can also love participating in activities or thinking about certain things. In a very real sense, we can experience love by engaging in the other virtuous pursuits described in this book. I love it when I get into a creative flow, when I contemplate some beautiful work of art, when I find some deeper meaning in an engaging novel, and when I feel valued as part of a community. Take a moment and think about something you love experiencing, doing, or thinking about. Maybe even close your eyes and try to visualize this activity or

DOI: 10.4324/9781003484783-7

experience. Perhaps you imagined sharing a bountiful meal with friends or family. Maybe you visualized relaxing on the couch with a special pet. As readers of this book, I suppose you might have even thought about spending some quiet time engaging in mathematically oriented puzzles or related activities. Whatever you pictured in your mind, the feeling that accompanied your activity or experience is so very special. It's special not only for the experience of it but also for the way that the experience remains with us. The ways that we give and experience love shapes our beings this love achieves a permanence within our minds and hearts. If the opportunity to develop an enduring sentiment of love within a math class strikes you as an ambitious direction for our pedagogical designs, I agree. This is very lofty, idealistic stuff. However, is teaching in a way that connects to love really that more audacious to imagine than creating a math lesson that centralizes freedom, beauty, or justice? Pursuits of love and permanence are another way that we, as humans, find fulfillment in our time together. Why shouldn't these virtuous pursuits intersect with our math teaching and learning? As Tina Turner once asked, "What's love got to do with it?" A great deal, I would argue.

I will admit, though, that turning the corner into practical matters, with respect to love and permanence, can be tricky. Love, in particular, feels like such personal and contextually situated experience, and one that would be hard to connect to our classroom teaching beyond a mere platitude of "loving math". Be that as it may, let me share one example from my own teaching that might provide a doorway into these rich and rewarding virtues. In this exchange, Amy and I (the Teacher) are working on something called a row task (Wright, Martland, & Stafford, 2006). Basically, this is a long strip of paper with a row of dots on it, and I used it to pose non-numerical arithmetic tasks where portions may be concealed or screened. Consider the exchange below (see Figure 7.1).

At first glance, aside from the novelty of the instructional materials, perhaps, this might not strike you as very related to topics of love and permanence. From my perspective, though, it's one of the first moments that I thought of when outlining this chapter. Up to this point, Amy had been very reliant on perceptual (i.e., concrete) materials to support her arithmetic reasoning.

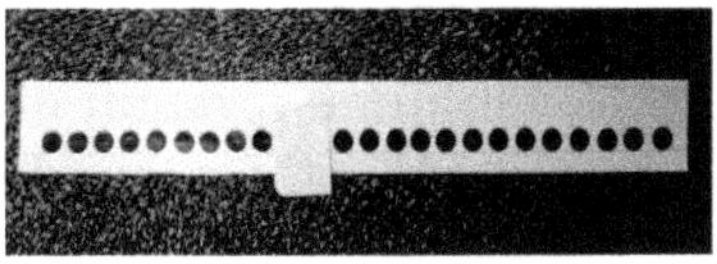

TEACHER: "Can you put that on number five?" [hands student a translucent blue counter]
AMY: [touches the first five dots on the tool] "one, two, three, four, five"
[places counter on the fifth dot]
TEACHER: "There's five."
TEACHER: [places a screen over four dots]
TEACHER: "This is ten." [places a translucent blue counter over the 10th dot]
TEACHER: "How many dots are under here?" [drags finger horizontally across screen]
AMY: [touches screen four times in a linear pattern] "Six, seven, eight, nine"
touches counter on top of 10th dot] "ten"
TEACHER: "Okay, so how many dots?" [taps screen]
AMY: "Four, you can't trick me."
TEACHER: "What were you thinking about when you were counting that?"
[taps screen four times in a linear pattern]
AMY: "I can see under it."
TEACHER: "You can see under it? Tell me about that. You can see the dots
through here?" [taps screen] "I can't see the dots through here." [taps screen]
AMY: "I can, I can see under it."
TEACHER: "So when you are pointing at that" [touches screen]
"you see a dot, that's what you're telling me."
AMY: "uh huh" [nods head]

FIGURE 7.1 Row Task with Amy

In particular, she really like using her fingers as perceptual replacements (Thomas & Tabor, 2012). When I would conceal or screen materials, she used her finger to replace whatever I had hidden. Of course, once the size of the hidden collection(s) exceeded ten, this strategy lost its viability. Thus, on this day with this task, I was so very pleased to see her touch and count imaginary dots that she couldn't physically see or feel. Her gentle taunt of, "you can't trick me" made me feel so celebratory in this moment. As she explained that she could "see under [the screen]", presumably within her own mind, the best word to describe my feeling was that of love. I loved that she had achieved the ability to construct a new and powerful type of unit – the figural (i.e., imaginary) unit (Steffe, 1992; Thomas & Harkness, 2013). I loved that we were sharing this moment of discovery together. I loved that my task design had worked as intended and that my work to better understand numeracy development had produced something real and powerful. Although I didn't ask her at the time, I suspect Amy loved unlocking a new capability with respect to tasks that had previously proven quite difficult or required her to enact a more limiting strategy. Mostly, I loved being a math teacher in that moment:

Take a moment to reflect on a time when you loved teaching math. Jot down your ideas here. Share this with your colleagues Ask them if they can think of a time when they loved teaching math.

____

____

____

____

It's worth mentioning, also, that this moment remains with me to this day. In that sense, it is permanent for me, and I can return to it, remember it, whenever I want to. Similarly, the strategy that Amy developed that day was permanent. It's not permanent in the sense that it she would never develop new and more efficient strategies to reason arithmetically, but that she would, from that moment on, be able to imagine and count hidden items. Of course, the researcher in me must make the caveat that, in this exchange, her strategy was likely still emerging, and she would continue to oscillate between perceptual and figurative counting for a brief while (Thomas & Harkness, 2019); however, generally speaking, Amy had developed a new skill that she would have in her toolbelt for life. Being able to see through that screen and count imaginary things was permanent for her.

These are but a few ways that we might operationalize love and permanence in our math classrooms. Indeed, this example is but a fairly cursory way into a conversation space that has remarkable depth and complexity. Permanence, for example, goes far beyond the persistence of an experience in our minds and into the ways in which humans and our ideas endure, even beyond lifetimes. Further, love, in this instance is, really, about me and Amy's experience of love is but a hypothesis. In the sections to follow, we will connect these ideas to specific teaching practices aimed at creating space for the pursuit of these virtues. Before we get into those practical matters, though, let's take some time to explore what we mean, exactly, by love and permanence.

The Virtue of Love

Love is, arguably, the greatest of the virtuous pursuits explored in this book. Perhaps rivaled only by play and meaning, love is so universally sought and experienced by humankind throughout our history. However, love addresses something deeper and more elemental than the other virtues in this book. It speaks to our desire to connect with those around us. To care for others and to be cared for ourselves. If community, as Aristotle suggested, is a body of individuals finding common purpose (Onwuatuegwu, 2020), then love is the blood that binds a community together.

If it still feels strange to use the word, love, to describe dynamics in the math classroom, I understand. Invoking love into our systems of education seems emotionalize them in a way that, unlike pursuits of virtues such as play or meaning, seems overly personal. With that in mind, let's consider the different ways we might invoke the idea of love. Interestingly, the ancient Greeks had several different words pertaining to different variants of love. I would guess that you are probably familiar with at least a few of these. *Eros,* referred to romantic or erotic love. *Storge* referred to the natural (or biological) love that a parent feels for a child and vice versa. *Philia* described the abiding love of friendship, and *Agape,* referred to an unconditional, limitless love that undergirded deeply altruistic acts (Soble, 1989). While not always using the specific Greek word, agape, the idea of selfless and infinite love is a crucial element of many religious traditions to describe the love that a higher power feels toward people and creation.

As teachers, perhaps the most germane of these ancient variants is philia or the love of friendship. Although, I freely admit that after some long days, our expressions of love may feel like they drift more toward the unendingly altruistic agape of the martyr. Nevertheless, the deep bonds that we form with our classroom communities are best described by a philial connection of loving friendship and care with those near to us. Both Plato and Aristotle described philia, or friendship, as a core element of human fulfillment (Murr, 2014). While we likely don't consider our students to be friends in the same way that we do our peers and colleagues, as teachers, it is undeniable that we form relationships with students and they with us. I don't know if I have ever met a teacher who didn't care deeply for the students in her/his/their classroom. Of course, sometimes our students can test our relationships, but in an interesting way, I have often found that I developed even stronger philial bonds with those students who presented social, emotional, or behavioral challenges. Finding our way through complicated interpersonal space is, somehow, strengthening to the love that I feel for these individuals.

In the math classroom, there are several different forms this philial love or friendship can take. In broad terms, we might categorize our love into two areas, love for our students as math thinkers and love of our students' math thinking and activity. The first, loving our students as math thinkers, refers to using mathematics as a conduit to more fully relate to those in our classrooms – "the love one might have for another human being through and because of mathematics" (Su, 2020, p. 205). In his chapter on love, Su describes how math provides a rich occasion to engage deeply with those around us, to talk with individuals about ideas, to explore spaces of anxiety, and to share our minds and hearts with one another. Su writes,

> To often, those of us who teach math professionally say, 'My job is to teach math', as if teaching math were only facts and procedures. We forget that 'my job is to teach people' whose experiences often interact with mathematics in completely different ways than our own experience do. And that means education has to take into account the whole person: the joys and sorrows of each person beyond just the mathematics they are learning . . . Because we are not mathematical machines, we breathe, we feel, we bleed. We are embodied human beings. Why

> should anyone learn mathematics if it doesn't connect deeply to some human desire, something we long for – to play, seek truth, pursue beauty, find meaning, fight for justice? You, as a mathematical explorer who is learning to embrace your identity in math, can be part of the movement to read people differently. *Believe that you and every person in your life can flourish in mathematics.* This is an act of love.
>
> (p. 208)

As math teachers, we have probably all heard the saying that everyone can be a math person. I believe this to be true very deeply. What Su suggests, though, is to live this vision in our classrooms, to lean into the human dynamics of doing math together, to believe in our students unconditionally, and by so doing, we are loving our students as mathematical beings.

Another way we might think about this virtue is the love we feel toward our students' thinking itself. On this point, a friend, scholar, and regular mathematics education collaborator, Beth MacDonald (2024), writes,

> Have you ever sat with a group of your students and wondered, 'What are they thinking?' and 'How did they solve that problem . . . In all my years, I still love listening to my students and young children's strategies and ideas. It gives me insight into their reasoning, point of engagement, noticings, questions, and wonderment . . . For me, by loving our students' mathematics, we examine the heart of mathematics teaching and gain new ways to think about mathematics teaching.
>
> (pp. 82–83)

What Beth is describing is the way that we may love, not only the mathematical beings that our students are becoming but also the ways that process manifests in their words, activity, products, and interactions. I will admit, this is sometimes challenging for me. Managing a math learning experience is complex work, and I often feel the pressing nature of time and the need to accomplish my learning goal for the day. Taking a beat to just enjoy and love the work of my students doesn't always feel very natural to me in the moment. However, when I do slow down a bit, spend an extra moment with a group or individual, focus more on questions rather than explanations, and just listen, it can be quite magical. Even in moments of struggle, my students often surprise me with novel thinking or just candor about where they are with a particular strategy. Those moments where I can be with them in the moment and engage, often leave me enthralled with their thinking. Remembering back to my work with Amy on the row task, when she told me that she could "see under [the screen]", I loved the way she phrased her thinking. I loved that it represented some real, sustained effort on her part, and that she was working at the cutting edge of her mathematical understanding. I loved it all.

Thus far, we have examined love mostly from the perspective of a teacher. Undoubtedly, this is important, as imbuing our practice with love is a rich and rewarding way to do this work. However, these ideas are also applicable to students. Most certainly, we can guide our students to see one another as mathematical beings. In addition to working through tasks as a small group, for example, we might also ask each group member to identify and share one thing the person to their right or left did that was interesting, helpful, or pushed the group further. Similarly, we can help our students love the mathematical thinking of their peers. One key practice of facilitating productive mathematical discussions is to connect students' strategies and ideas (Smith & Stein, 2018). This practice involves drawing attention to the uniqueness and worth of ideas and helping students to engage with one another's thinking from a place of appreciation. In short, in addition to students' considering the math of their peers, we can help them see goodness in that thinking – to find love in the words and activity around them. Core to these processes is the formation of relationships. One way or another, we will develop some manner of relationship with our students as we share space and time together, and they will develop a relationship with us and their peers. My goal is to create a culture such that these relationships embody a philial love for the ways we may apply our minds to math tasks as well as the mathematical beings we are becoming.

The Virtue of Permanence

At this point, I would imagine that most of the virtues we have examined are somewhat familiar. Likely each of us has some experience contemplating, seeking, or experiencing beauty, justice, truth, and so forth. If you are like me, though, permanence may seem a bit more slippery as a concept. What do we mean by permanence and why would this be virtuous or contribute to human fulfillment? In short, permanence refers to the way that things, ideas, emotions, memories endure among us. In mathematics, we can use lasting consistency of foundational principles to build interesting and elaborate skyscrapers of ideas. More broadly, there is often a deep fulfillment that comes from the legacy we leave via our lived experiences and memories. My family and I spend a lot of time at Lake Cumberland located in the southern portion of Kentucky. I find great joy in the notion that the memories we create there every summer will be held by my children, and perhaps someday be passed to their children in an oral tradition. So many stories of old from my parents, grandparents, aunts, and uncles inhabit my mind to this day, and we return to the more outlandish ones with some regularity. There was the time that a distant cousin of a great grandparent, who went by the deliciously Appalachian handle, "Mooney Lightning" came down from the hills and proceeded to shoot holes in the family homestead with his pistol just for fun. There were the Christmases when (during the great depression) a giant stalk of bananas would be purchased and hung in the living room for young people to enjoy – bananas in Kentucky (during winter, no less) were quite a treat in those days. In the same way, my hope is that this legacy of summer lake fun will enjoy a sort of permanence across generations with stories about the night swims, the lakeside fireworks, and so on persisting through time. As human beings, we find delight, utility and fulfillment in the things that persist and endure. But don't just take my word for it, permanence, a virtuous pursuit, has been with humanity for quite a long time.

Looking back through time, once again, we might begin with our ancient Greek guide, Aristotle; however, contemplations of permanence (like other virtues) is something that is shared across time, culture, and geography. Nevertheless, let's consider a few terms offered by Aristotle and contemporaries to shed some nuance on this notion of permanence. Fundamentally, for something to be permanent, it must remain in existence in some way, but how? Ancient philosophers used the term *endurance* to describe things that are "wholly present at each time at which they exist, so that they have no proper temporal parts" and *perdurance* to describe things "composed of temporal parts, generally point-in-time parts" (Pemberton, 2022, p. 217). In other words, one way to think about permanence is to consider how things endure through time. The spoons in my kitchen drawer, provided I don't use them, remain the same spoons day after day – they endure as spoons through time. Such philosophers also used the term perdurance to describe the way that time inflects our experience of permanence. From a perduring perspective, I am a person whose existence is braided with the passage of time. I have existed as a child, an adolescent, and an adult, and each of these life seasons marks my existence in a very temporal, or time-bound, way. Aristotle considered the ideas of endurance and perdurance in concert to create a view of permanence that Pemberton referred to as "lasting". Specifically, what Aristotle sought was a way to describe how something might move through and be shaped by time but also remain the same throughout time – to be permanent. As a human, I am perduring in the sense that I am comprised of temporal eras, but I also endure as the same person throughout these eras. In that sense, for my time on earth, I am lasting. In the same way, undoubtedly those stories of old that I shared have been shaped by the tellings and retellings over the years, with different versions existing in a given era, but there is a fundamental sameness to Mooney Lightning and the Christmas bananas that endures. These memories last for as long as they continue to be shared. Moreover, it's not just memories that may last; ideas last, too. The virtues we have explored in this book are all lasting in the sense that humans across millennia have sought to understand and experience them. Love, for example, is permanent to the human condition. By that, I don't mean that all human existence experiences a constant state of love, but that love is an inextricable element to being human that lasts throughout time. Love moves with us as people. It is an idea, feeling, and pursuit that travels alongside and within us. Love lasts.

In mathematics, permanence holds a special meaning. In his book, *A Treatise of Algebra* (1842), George Peacock described a principle of permanence that described the ways that operations should behave consistently across different conditions and number systems. Looking further back, we can find this notion of permanence in the postulates of Euclid's *Elements* (1956 trans.) – that a line exists between two points and that all right angles are equal to one another, for example. These are lasting ideas in the sense that they endure consistently through time, but also perdure in the sense that mathematical thinkers of different eras have considered them, applied them, and built upon them in unique ways. In that sense, mathematical permanence is simultaneously timeless and moving through time.

Looking beyond the nuanced temporal aspects of such permanence, though, the ways in which these ideas last can be a source of deep fulfillment and power. On this point, Su (2020) writes,

> We adore *constants*. The word *constant* is often used to describe some fixed number that has significance, like the golden ratio, e, or π. The number π captures the the public imagination in part because it arises in many places, not just the geometry of circles. Some instances are quite unexpected, such as the sum of reciprocals of square numbers.
>
> $$1/1^2 + 1/2^2 + 1/3^2 + \cdots = \pi^2/6$$
>
> or in the formula for the area under a bell curve, or in Heisenberg's uncertainty principle. As such, the number

> π feels cosmic – mysterious and eternal, no doubt and important constant at any point in time and in any corner of the universe. It is no surprise that some people try to grasp this number's inscrutable nature by memorizing its digits or seeking patterns in them. We also search for *invariants.* In mathematics, an invariant is something that remains unchanged when you perform an operation. For instance, if I multiply a number by 5, that doesn't change whether it is even or odd. If I rotate a 3-D geometric figure, that won't change its volume. Invariants reveal insights about the operation itself. For example, I can learn something about a rotation by what doesn't change: the axis about which the rotation spins is fixed and reveals an important feature of the rotation. A key to solving a Rubik's cube is paying attention to what doesn't move when you make a move. In mathematical models of physical systems, a conservation law focuses on an invariant quality that doesn't change as the system evolves – such as the total momentum when two cars collide.
>
> (pp. 89–90)

What Su describes here is the way that mathematics itself provides us with a place of lasting permanence. Somewhat similar to the way a memory remains permanent in is passage through time via storytelling, constants, invariants, axioms, and so forth remain permanent places where we can gain sure mathematical footing for our thinking. Unlike memories, though, these mathematical structures enjoy a special exactitude that makes them particularly enthralling. Depending on who is telling the story, Mooney Lightning may have shot up the walls of the homestead 5 times or 50 times, but π is always π. We might say that it's perfectly permanent in the way that it lasts with such constancy. To the extent that we may connect our classrooms to ideas, experiences, truths, and histories that last, we tap into some deep human pursuit of the permanent. Su (2020) writes,

> Trust is at the heart of a desire for permanence. If I seek a refuge, it is because I need a hiding place to be safe.

> If I look to a yardstick, it is because I expect it will not change. If I step in a foothold, I must have confidence that it will hold me up.
>
> (p. 97)

Fostering a sense of permanence in our classrooms is to build trust in our own mathematical ideas and in the people and the world around us.

Core Practice: Beginning with Love in Mind

I did my student teaching just outside of Lexington in the small Kentucky community of Versailles (pronounced, in true rural fashion, Vuhr-Sayles not Vuhr-Sy). My cooperating teacher, Ms. James, was a remarkable educator and one of her favorite sayings was to "begin with the end in mind". That is, she encouraged me to identify what I hoped to accomplish in any given lesson; what standard or learning target I aimed to meet and work backward from there with my planning. Her words remain with me today, only now they seem to encompass so much more than just standards and lesson objectives. If my goal is to engender love in the math classroom, I begin by imagining what that love might look like and work backward from there, or, to paraphrase Ms. James, I begin with love in mind.

Ms. James also showed me something else as she guided me toward becoming a teacher. Hers was a fifth grade classroom, and she reserved a portion of each day for students to find a place away from their desk (a comfy mat or beanbag chair, etc.) to read something of their own choice. This was a neat bit of pedagogical strategy in that it allowed Ms. James to catch up on tasks or prepare for the lesson to come. However, Ms. James also made sure to engage with the moment itself. She would take note of what books her students had chosen to read and ask them about those books. She seemed to deeply care about why a particular character made one decision over another or the details of a fictional landscape in which a story took place. One day, during this time, you could hear the faint murmur of whispers from more than a few students as they read, mostly, to themselves. Ms. James nudged me on the shoulder and said, "Listen to that. It's like music." In that moment, I could see how Ms. James had fallen in love with her work. How she reveled in the sights and sounds of learning and how I could do likewise. In a literacy space, Ms. James was experiencing the same love described by Beth MacDonald earlier in this chapter and she had done so with intentionality. She created spaces where love could exist and endure in her classroom. Since that moment, whenever I hear young people reading (nearly) silently to themselves, I hear that lovely music yet again.

Of course, this experience of love in the classroom is from my perspective as a teacher. But, in a very real way, this is where the core practice of beginning with love in mind begins. The first step is to open up space for us, as teachers, to experience love. That is, we need to acclimate ourselves to loving our students as mathematical beings and loving their ideas and activities in our classrooms. We listen for that music and appreciate it so that we can help students feel such things about themselves and others. Beginning with love in mind, my goals are a) for students to experience some real attraction to the mathematical thinking and activities of their classmates and b) to see one another as valuable and capable individuals who belong and contribute within math spaces.

Earlier in this chapter, I made a brief reference to practices aimed at facilitating effective mathematical discussions. Smith and Stein (2018) describe a sequence of five practices that teachers may enact to maximize student discourse around a mathematical topic. First, we *anticipate,* and this involves imagining how our students will respond to the tasks we pose. What are the different directions their thinking may take them? Then, we *monitor* which involves watching and listening to our students closely as they engage in the mathematical activity. We can think of this as professional noticing in the moment. Next, we *select* particular examples of students' thinking that we feel would advance the math conversation and move us toward our goal. Then, we *sequence* those selections. We consider what is the best order for our selected students or student groups to share such that we create a coherent path toward our goal. Lastly, we *connect* these shares strategies in a way that brings key ideas and concepts to the forefront.

Of particular interest, regarding the notion of love and permanence, is the practice of anticipation. When thinking about the varying strategies students might bring to bear on a task, we can envision the energy and effort they expend to think in these different directions. For each different pathway you imagine students might take, try to find something good within – something to love about it. Even anticipated strategies that may seem unproductive or rudimentary in some fashion, likely there a core goodness that we can identify, elevate, and love. A few times in this book, I have referred to the *believing game* (Elbow, 2006; Harkness, 2008). Recall that believing, in this context, refers to our trying willingness to search for truth in students' thinking, and to dial back our own sense of mathematics a bit to *believe* in what our students share with us. When it comes to anticipating opportunities for love in a mathematical moment, we go one step further with our belief. We search for ways to love their thinking. We search for goodness in their ideas and imagine how we might celebrate that goodness explicitly among our students. The goal is to model the practice of falling in love with the ideas in our classrooms so that our students may see this in action, and it begins with *anticipating* where those intersections of love and thinking might occur in our math moments and *believing* in some core goodness of each students' ideas. For example, I anticipated that Amy (see Figure 7.1) could construct imaginary dots in the place of those hidden by a screen on that row task. In the grand scheme of unit construction, this is still some distance from the facile foundational numeracy associated with creating and using multiple composite units (Thomas & Tabor, 2012). However, watching a student who has, thus, far relied on physical items to take that step toward interiorizing mathematical units is quite thrilling. Not only can I anticipate follow-up tasks and questions, but I can also imagine the ways that I might show my love for her thinking in that moment. Thinking back, I didn't say then what I would like to say now. This was a missed opportunity for me to bring the virtue of love into this interaction. If I could travel back in time, I would say something like, *"My goodness, I am amazed at how you are able to see under that cover in your mind. That is so special, and I love that!"* I would want to explicitly lift up Amy's thinking and share my love for it. I want to model, for her, how to love someone's math thinking and then give room for her in another lesson with peers to do likewise. With each opportunity to experience the love for another's ideas and place as a mathematical being, the practice moves toward permanence – it becomes a lasting approach for our community.

If this sounds a bit simplistic and trite, perhaps it is. As the Beatles sang, *"All you need is love"*, but to take it seriously in our teaching is quite another thing. Finding ways to love our students' as mathematical thinkers and their thinking is, fundamentally, an asset-based approach to teaching (Celedón-Pattichis et al., 2018). That is, we aim to consciously "move away from deficit perspectives that view students, parents, and communities as lacking" (p. 375). Rather, we work to valorize the different ways that students use their minds, the funds of knowledge they bring to the table, and contexts that shape students' activities (Civil, 2017). We seek to find something good in students' thinking upon which we may build. In my experience, though, this is often easier said than done. So much of our educational landscape and language is built around the notion of misconception. Where did my students falter in their thinking and how do I fix it? Terms like remediation and intervention establish a mindset of looking for ways our students are falling short. I have spent a great deal of time thinking about what my students are missing and how distant their thinking is from my intended math goals that I lose sight of the amazing things they are doing in the moment. Rather than seeking to repair some misconception, I look for some wonderful foundational thinking upon which we can build. It may seem like a subtle shift in outlook, but there is much in our profession that steers us in the opposite direction. Beginning with love in mind means to find the places where goodness may emerge, where we may fall in love with our students' thinking and activities, and where we bring that experience of love to the forefront so that our students may do likewise.

Connecting Practice: Professional Noticing for Love and Permanence

If the preceding section still seems a little heavy on platitude, let's operationalize love and permanence in the math classroom a bit more practically. For those who have read from the beginning, likely you are well familiar with the practice of professional noticing by this point. However, for readers who may have jumped directly to this chapter, professional noticing describes a way of teaching mindfully and responsively. It is composed of three component processes. Attending, which is closely watching and listening within a mathematical moment to gather as much detail as possible. Next, we interpret what we have seen and heard to develop some sense of what our students understand about the mathematics at hand. Lastly, based on our interpretation, we make some decision aimed at advancing students' knowledge and practices which moves us closer to our mathematical goal.

To delve into professional noticing for love and permanence, let's begin by considering a particular math goal related to fractions. In my experience, fractions are one of those mathematical places where, to borrow a phrase from Su (2020) students' confidence in their own permanent mathematical footholds erodes a bit. Fractions precipitate a degree of units construction and coordination across multiple levels that can often be quite challenging for children (and adults). Consider the following fifth grade standard,

> Solve word problems involving fraction arithmetic referring to the same whole, including cases of unlike denominators, e.g., by using visual fraction models or equations to represent the problem. Use benchmark fractions and number sense of fractions to estimate mentally and assess the reasonableness of answers.
>
> (NGA & CCSSO, 2010 – see standard 5.NF.2)

This is one of those standards a teaching friend of mine refers to as *a whopper*. That is, there is a lot going in this one. First, we need to think about word problems or tasks in context. Then, we need to make sure our tasks involve specific forms of fraction arithmetic. Also, we want to create space for representations and models. Lastly, we want to have opportunities for students to think metacognitively about their own math processes. Challenge accepted! A while back, some colleagues (including my mentor, Shelly Harkness) and I engaged in a lesson development experience aimed at this standard (Harkness et al., 2013; Lane, Thomas, & Harkness, 2013). Our goal was to design a really strong plan and sequence of tasks, and we landed on the idea of cornhole.

For those who may not be familiar, cornhole (sometimes called "bags") is a game played with two boards approximately 2′ × 4′ (see Figure 7.2) and beanbags. Each board has a 6″ diameter hole near the top. The two boards are set approximately 20′–30′ apart (27′ is competition regulation). Players stand beside each board and take turns throwing four bean bags each into the opposing board.

If you throw your beanbag into the hole on the opposing board, that's worth 3 points. If your bag lands (and remains) on the board but doesn't go in the hole, that's worth 1 point. Any

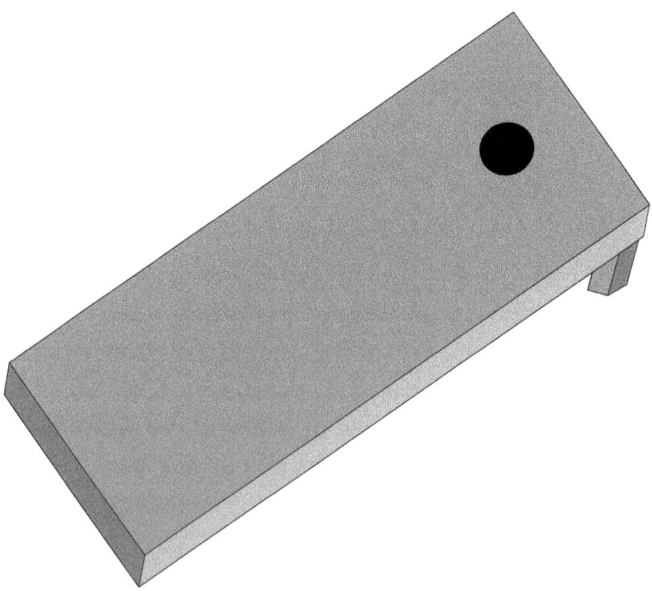

FIGURE 7.2 Cornhole Board

other throw is not worth any points. One other note, cornhole boards are often painted or decorated in some manner – favorite sports team logos are particularly popular.

Over dinner, one evening, my friends and I began thinking about a scenario where the size of the cornhole board might vary. Because one earns points for landing a bean bag on the board, a larger board would make the game easier for novice players. Similarly, a smaller board would prove more challenging. Then, we thought about how varying the board size might impact the amount of paint needed to decorate each board. As our thinking developed, we envisioned a series of tasks around this context (see Figure 7.3).

We then proceeded to teach and video record the lesson in various classrooms with the goal of refining the tasks and our questioning process such that we developed the strongest instructional experience possible (see Harkness et al., 2013 for a complete account). The real key to the lesson, though, was the

CONTEXT

Because we are not making enough money as a teachers, we have invented a variation of corn hole that has three skill levels.

The boards come in small, medium, and large. The medium board is the standard corn hole size. The large board is for the novices. The small board is for the experts.

TASK 2

We have received an order for 4 large boards for the kids.

This is a way to increase their self-esteem. One quart of paint will cover 2/7 of a large board.

We are being really stingy with my paint, so how much paint will I use to paint the boards?

Draw a picture to represent this problem and your solution.

TASK 1

Times are tough, so we do not want to purchase or waste any extra paint.

Each quart of red paint covers 1/2 of a medium board.

If we have 4 medium boards, how much paint will we use to cover all of the boards?

Draw a picture to represent this problem and your solution.

STUDENT WORK – TASK 2

FIGURE 7.3 Cornhole Lesson Tasks 1 and 2

emphasis on representations. We explicitly guided students to use their drawings (rather than bare numbers or equations) to support their solutions. The first two tasks were designed to be a relatively accessible entry point, and for the most part, students didn't experience too much disequilibrium. Often, at first glance, the "large board" task where a quart of paint only covers 2/7 of a board seemed challenging, but fairly quickly, student groups were able to devise a representation. While these representations varied a bit, typically they appeared as rectangles (cornhole boards) divided into sections that matched with the task (e.g., halves and sevenths) that were then colored in. *Take a moment to draw your own representations for tasks 1 and 2.*

TASK 1	TASK 2
4 medium boards	4 large boards
Quart of paint covers 1/2 of a board	Quart of paint covers 2/7 of a board
How much paint will you use?	How much paint will you use?

Task 3, though, was designed to be somewhat more challenging. Imagine that we want to paint four small boards. In this case, a quart of paint would cover nearly (but not quite) an entire board. Specifically, it would cover 7/8 of a board. *Take a moment to draw your own representations for task 3.*

TASK 3
4 small boards
Quart of paint covers 7/8 of a board
How much paint will you use?

If you are like most folks (including other adults) who have worked on this task, you probably drew four cornhole boards and divided each board into eight sections. Coloring in seven of the eight sections on each board would represent four quarts of paint. However, there is still the lingering single "unpainted" section on each board to contend with. In total, there are four of those unpainted sections. How much of a quart of paint does that represent? In my experience leading this lesson many times, most students find themselves in a conundrum and their thinking oscillates between 4/8 (or half) of a quart or 4/7 of a quart. Because they have divided their board into eight sections, the remaining four unpainted sections appear, so naturally, as a "half" that it's challenging to escape that conclusion (half of a quart). However, the problem itself, undermines that conclusion – if a half a quart covered half of a board then shouldn't a whole quart cover a whole board? But in this case, it only covers 7/8 of a board.

I want to pause for a moment and share how much I absolutely love everything about this lesson. I loved the process of working with dear teaching colleagues to think through an engaging and challenging task sequence. This form of pedagogical engineering is one of my most favorite aspects of being a teacher. I loved sharing my experiences leading this lesson with my colleagues, watching videos of students, and thinking hard about great questions we might pose in the moment, missed opportunities, and what math might come next. Mostly, though, I loved the way that my students, when they worked on task 3, would dig in and embrace disequilibrium, seek to understand the thinking of their peers, and work to resolve mathematical tensions through productive discussion. In one particular class, about half of the student groups were convinced that it would take 4 and 4/8 (or half) quarts to complete the task, while the remaining groups were certain that it would take 4 and 4/7 quarts. One of the 4 and 4/7 groups decided to cut out a small rectangle to represent the remaining paint and use that to support their argument. Consider the following representation and exchange (see Figures 7.4 and 7.5).

TASK 3	STUDENT WORK – TASK 3
Word is getting out and our business is picking up. We have received an order for 4 small boards. **This is the really challenging board. One quart of paint will cover 7/8 of a small board.** **We are trying to make sure that we keep my profits high, so we are still not wasting any paint. Now, how much paint will we use to paint the boards?** **Draw a picture to represent this problem and your solution.**	

FIGURE 7.4 Cornhole Lesson Task 3

TEACHER: *"OK, so I see you all landed at 4 and 4/7, but you all have another part to your drawing. Tell us about that part."*

ALYSSA: *"Yeah, it was really hard to hold onto the 4/7, because it looks like a half. Like, it's 4/8 of the board, and Zoe kept arguing for 4/8. But you have to think about quarts so we tried to make something that would make us think about quarts."*

TEACHER: *"What do you mean by think about quarts?"*

JADA: *"Quarts are sevens, I mean sevenths. Each quart is seven slices, so you have to think about how much of the seven slices you need. We needed four of the seven slices, so it's 4/7. The paper is so we could help us see that 4/8 of the board is actually 4/7 of the quart"*

TEACHER: *"Wait, 4/8 is the same as 4/7? Is that right? How does that work?"*

ZOE: *"It sounds weird, but it's like two different wholes. There is 4/8 of a board left to pain, but it's only going to take 4/7 of a quart. You have to keep thinking about boards and quarts as different things, and that's what is hard on this one. That's why we made the little paper thing."*

TEACHER: *"What do other groups think about this. Maybe somebody from a 4/8 group?"*

SELWYN: *"I really like the paper thing. I think I get it."*

TEACHER: *"Say more on this, Selwyn. What do you like about the paper thing? What makes it good?"*

SELWYN: *"I love how you can move it around. Like, it helps me see the quart better. It still kind of feels like it should be 4/8 to me, but I think it's 4/7. A can of paint is seven slices, and we only need four like on that orange paper. I like what y'all* [Alyssa/Jada/Zoe] *did there."*

FIGURE 7.5 Cornhole Lesson Task 3 Exchange

Digging into our practice of professional noticing, we might first *attend* to the within-group disequilibrium between the two positions (4 and 4/8; 4 and 4/7) as well as the way members of the group worked to resolve that tension – by creating a supplemental material to better represent their solution. *Interpreting* their activity, we might determine not only that this augmented representation adds power to their thinking about differing wholes (e.g., quarts, boards) but also illuminates their propensity to innovate – to think outside the lines a bit and bring additional, movable, elements to their static drawing. When it came to *deciding*, the teacher deliberately seeks input from other groups and, hearing Selwyn's affirmation, probes to better define "what makes it good". In this moment, we can see the combination of some very rich and complex mathematical reasoning, but also the ways in which students can express their affinity (or love, in other words) for the thinking of their peers – "I love how you can move it around". Moreover, these students' discussion around a unique representation helped Selwyn's sense of fractions solidify and become more permanent. His mathematical foothold got stronger as he considered the "paper thing". As teachers, we can model these asset-based practices and then create space for our students to do likewise. Two key things to keep in mind when noticing for love and permanence are to, first, model our appreciation and love for how our students engage with the math at hand, and lift the goodness in their thinking rather than foreground the misconceptions and shortfalls. Second, we want to create space for our students to do this, too. We want to help them affirm their peers and share openly about the interesting ideas of others and how they, perhaps, help one's own thinking come into better focus. When we endeavor to see our students as mathematically capable individuals brimming with remarkable ideas, when we love them and their thinking openly, they begin to do so themselves.

Questions for Reflection

Once again, as we near the end of this examination of love and permanence in our math classrooms and communities, I would like to offer some questions for your reflection. I encourage you to spend some time with them individually and then, perhaps, share your thoughts with a friend or colleague and gather that person's thoughts. While mathematical permanence may seem like a bit more familiar territory, the concept and expression of love in our schooling settings may strike you as a bit more distant from the work that we do. Hopefully, this chapter has helped illuminate practical ways that these virtues may inflect and influence our math teaching and the experiences of our students. As with the reflective questions in preceding chapters, the ones I pose below may not be easily answered, but my aim is to provide you with space to ponder a bit, to find the ways, where love and permanence may manifest in our work.

- *Do you think that expressions of love as described in this chapter (e.g., love of students as mathematical beings, love of students' thinking) are an important part (or should be an important part) of the learning experience? If so, why? If not, why not? Do you think others in your broader community would agree?*
- *What mathematical ideas and concepts hold the most permanence with you? Where do you feel most mathematically sure-footed, and how did this feeling develop? What led to your trust in these ideas?*
- *What do you love about teaching math? How often do you get to experience this love in your work? If you experience love in your work often, why do you think that is so. If less often, what might you change or work toward to experience love more frequently?*
- *How would you describe the extent to which your students develop permanence with their mathematical ideas? Do you feel like your students often develop lasting foundations that stay with them over time? If so, what do you do that contributes to this? If not, what prevents you from helping students achieve such permanence?*
- *What would you say to people who might not agree that love and permanence are important virtues to connect with our math teaching? What do you imagine they might say back in response? What are the challenges to bringing these ideas into a math classroom?*

References

Celedón-Pattichis, S., Borden, L. L., Pape, S. J., Clements, D. H., Peters, S. A., Males, J. R., . . . & Leonard, J. (2018). Asset-based approaches to equitable mathematics education research and practice. *Journal for Research in Mathematics Education*, *49*(4), 373–389.

Civil, M. (2017). Preface. In S. Celedón-Pattichis, D. Y. White, & M. Civil (Eds.), *Access and equity: Promoting high quality mathematics in pre-K–Grade 2* (pp. v–viii). Reston, VA: National Council of Teachers of Mathematics.

Elbow, P. (2006). The believing game and how to make conflicting opinions more fruitful. In C. Weber (Ed.), *Nurturing the peacemakers in our students: A guide to teaching peace, empathy, and understanding* (pp. 13–25). Portsmouth: Heinemann.

Euclid. (1956). *The thirteen books of Euclid's elements* (2nd ed., T. L. Heath, translator). New York: Dover Publications.

Harkness, S. S. (2008). Social constructivism and the *Believing Game*: A mathematics teacher's practice and its implications. *Educational Studies in Mathematics, 70*(3), 243–258.

Harkness, S. S., Thomas, J., Lane, C., & Cooper, A. (2013). Lesson study: Allowing "What is the whole?" to usurp "Where is the reciprocal?" *Far East Journal of Mathematics Education, 10,* 1–30.

Lane, C., Thomas, J., & Harkness, S. S. (2013). What is the whole in cornhole? Introducing and capitalizing upon disequilibrium with fraction operations. *Ohio Journal of School Mathematics, 67,* 33–41.

MacDonald, B. L. (2024). Loving our students' mathematics. *Mathematics Teacher: Learning and Teaching PK-12*, *117*(2), 82–83.

Murr, D. E. (2014). Philia in Plato. In G. Gurtler & S. Stern-Gillet (Eds.), *Ancient and medieval concepts of friendship* (pp. 3–34). Albany: State University of New York.

National Governors Association (NGA) Center for Best Practices & Council of Chief State School Officers (CCSSO). (2010). *Common core state standards for mathematics.* Washington, DC: Authors.

Onwuatuegwu, I. N. (2020). Concept of community: Aristotle's perspective. *International Journal of Research and Scientific Innovation*, *7*(7), 60–64.

Peacock, G. (1842). *A treatise on Algebra*. London: Cambridge University Press.

Pemberton, J. M. (2022). Aristotle's alternative to enduring and perduring: Lasting. *Ancient Philosophy Today: Dialogoi*, *4*(2), 217–236.

Smith, M., & Stein, M. K. (2018). *5 practices for orchestrating productive mathematics discussion*. Reston, VA: National Council of Teachers of Mathematics.

Soble, A. (1989). *Eros, agape, and philia: Readings in the philosophy of love*. New York: Paragon House.

Steffe, L. (1992). Learning stages in the construction of the number sequence. In J. Bideaud, C. Meljac, & J. Fischer (Eds.), *Pathways to number: Children's developing numerical abilities* (pp. 83–88). Mahwah, NJ: Lawrence Erlbaum.

Su, F. (2020). *Mathematics for human flourishing*. New Haven, CT: Yale University Press.

Thomas, J., & Tabor, P. D. (2012). Developing quantitative mental imagery. *Teaching Children Mathematics, 19*(3), 174–183.

Thomas, J., & Harkness, S. S. (2013). Implications for intervention: Categorizing the quantitative mental imagery of children. *Mathematics Education Research Journal, 25*(2), 231–256.

Thomas, J., & Harkness, S. S. (2019). Tethering towards number: Coordinating cognitive variability and stage-oriented development in children's arithmetic thinking. *Mathematics Education Research Journal*, 31(3), 325–347.

Wright, R. J., Martland, J., & Stafford, A. (2006). *Early numeracy: Assessment for teaching and intervention* (2nd ed.). London: Paul Chapman Publications/Sage.

8

Conclusion

For each chapter, I have curated a playlist from my own music library. These are songs that put me in the mindset of the topics at hand. Some may be on the nose and others may feel like a bit of a stretch. Hopefully, though, you find the music an enjoyable addition to your thinking in these areas.

Note: Spotify automatically adds selections to a given playlist. My original selections are the first 6–8 songs on a given list

https://open.spotify.com/playlist/5Wt1ONaJ5h1mjxJcszeXqZ?si=vniY-RaVT_qAglD1Y3qnMw

As we move toward the conclusion of our time together with these ideas, my hope is that you have developed a better sense of how different virtues may manifest in our math classrooms. My goal is to present some possible pathways and ideas based on my own work and experiences in this space. I imagine that not every example, suggestion, or task is directly applicable to your teaching practice. Perhaps some are, and that is wonderful. However, I hope that those examples and tasks might provide you with a platform to dream and vision things that may work in your classroom, with your students, at your school. One of my favorite things about teaching math, something I love, is taking an interesting task and modifying it to further some mathematical

DOI: 10.4324/9781003484783-8

goal and meet the needs of my students. I worry, at times, that space for such pedagogical engineering is becoming increasingly scarce, and I wholeheartedly invite you to keep that work alive and well. The flow of curricula, assessments, and other instructional resources through teachers is unavoidable. The only question for us is what we do with these tools we have given. Do we act as a more passive conduit, or do we engage with our resources and consider how to amplify them such that they bring out the most rewarding forms of math teaching and learning? I choose the latter, and I hope this book helps you imagine how to do likewise.

In this book, we have examined several virtues (e.g., play, perseverance, meaning, and truth). While I find that the pursuits of these particular virtues to have special connection to our work as teachers, there are several others that we could have explored together. One of the many things that makes being a human being wonderful is that there are so many forms of goodness we may pursue, and so many different roads we might travel to find fulfillment. Let's take a bit of time to consider some of the roads not taken in the preceding chapters.

Responsibility is most certainly a virtuous pursuit and one that has direct import for math teaching and learning. Finding ways to be responsible allows us to find a connection with the world around us. Being a parent, teacher, and pet-owner, the responsibilities I hold in those spaces create a deep sense of value. I am invested. I am empowered. Being responsible moves me away from being self-centered and toward being other-centered. Like most pursuits of virtue, responsibility requires energy. Ultimately, though, commitment to being responsible is another way of moving toward fulfillment and flourishing.

Fostering a sense of responsibility for one's own mathematical participation is crucial to the learning experience; moreover, being a responsible member of a math community certainly enhances the ways students work with one another and engage in meaningful activity. In terms of our teaching, we may leverage some of the ideas discussed in our exploration of community and perseverance to also connect with responsibility. For example, when we co-create definitions of community with our students, we might introduce responsibility to one another as a guiding practice for participating in such communities. This might include committing to the work at hand (i.e., being responsible for our own participation) and engaging productively with our groups (i.e., being responsible to our immediate community). In any event, responsibility is an example of a virtuous pursuit that we might have explored more deeply but didn't.

There are many virtues, kindness, and courage, for example, that we also might have investigated and connected to our teaching practice. Fostering a sense of kindness, care, and empathy for those around us is a deeply fulfilling approach to living life. Similarly, finding moments to exercise courage can also be quite rewarding. Stepping into the unfamiliar, trying something new, and voicing one's ideas, even when tentative, are all courageous forms of mathematical participation and provide a foundation for enhancing confidence in mathematical spaces. There are many other virtuous pursuits that might inflect our mathematics teaching, and I wholeheartedly encourage you to seek them out. My hope is that this book has given you some initial sense of what connecting virtue to mathematics teaching and learning looks like, such that you may expand your field of vision to many other ways we might rehumanize math and create spaces for student flourishing.

Virtue, Mathematics, and Current Trends

Sometimes, when I read a book or article from decades prior, there are contextual factors that date the references in some way. In a few places, throughout this book, I make mention of the Common Core State Standards for Mathematics (NGA & CCSSO, 2010); however, since the inception of those standards in 2010 (and subsequent adoption by 46 states), the initiative has waned with several states (including Kentucky) repealing their initial adoptions. Not that long ago, Massive Open Online Courses (MOOCs) offered by universities and enrolling thousands of students per section were put forth by many in higher education as a productive disrupter which would fundamentally change the university landscape. Spoiler alert: MOOCs have pretty much left the scene. Years ago, the radio and the television were both envisioned as technologies that would profoundly change schooling. Similar to what many of us experienced in the COVID-19 pandemic and rise of distance/virtual teaching (e.g., Zoom classrooms and lessons), the polio epidemic of 1937 prompted Chicago (and other large cities) to deploy "radio schools" where curricula were broadcast to students over the airwaves (Foss, 2020). As you might imagine, this prompted scholars and policymakers to envision where such broadcasting tools might take the educational experience. Ultimately, though, these technologies were incorporated into schooling, but did not result in some fundamental restructuring envisioned by many.

On the other hand, sometimes technologies and ideas do emerge that profoundly reshape how we conduct education. The advent of the internet, in my estimation, is one such technology. From the ways that students engage with media, to facilitating distance interactions, conducting assessment at scale, not to mention the complicated interpersonal and communicative dynamics introduced by social media applications, the internet has had a substantive and lasting impact on teaching and learning at most every level. I mention all of this because the next topics we will explore will likely date this book in one manner or another – specifically, Diversity, Equity, and Inclusion (or DEI) and Artificial Intelligence (AI). As I write this, both DEI and AI are at the forefront of many educational conversations, and there are many ways they intersect with the ideas of this book. Certainly, there is a pathway where DEI and AI recede from our collective consciousness (a la the Common Core State Standards and MOOCs). Alternatively, our discussion here is prescient with respect to a future educational landscape where these DEI and AI play an increasing role in shaping the nature of schooling. Either way, given the current amount of energy in this area, I find it useful to consider how virtue and human flourishing in the math classroom intersect with these topics and technologies.

Diversity Equity and Inclusion: What's the Big Deal?

At risk of drifting too far into editorializing, my sense of the discourse around DEI is that it is often ill-structured and evanescent. Terms like "woke" and "CRT" (which stands for Critical Race Theory) sometimes emerge in seemingly synonymous ways, which can make it harder to discern what, exactly, we are talking about. With that in mind, let's take each letter of the acronym and map it to virtues and practices discussed thus far. "D" stands for diversity. In the context of a math classroom, diversity refers to the presence of difference; different lived experiences, different belief systems, different levels of capability, different approaches to problem-solving, and so forth all coexisting to form a rich, heterogenous community across many domains. Proponents of DEI portray diversity as a strengthening axis for schools (and math classrooms by extension). As a teacher, I tend to agree with this portrayal. The types of conversations I can facilitate among students who have different backgrounds and different ways of thinking are far more interesting and productive than those where everyone brings the same things to the table and has the same approach to their math thinking. It's analogous to visiting Baskin-Robbins and only finding vanilla or chocolate ice cream – I want 31 flavors from which to choose! Rich math activity and discussion is implicitly predicated on the presence of diverse perspectives and ideas. Rather than being drag on educational processes, diversity is, in my estimation, essential to teaching math for human flourishing. Diversity in varying ways allows for a more elaborated pursuit of meaning and opens up spaces for different forms of beauty, just to name a couple of connections. In a landscape of increasing curricular homogenization and persistent racial, ethnic, and class divides via residential segregation (Charles, 2003; Denton, 1995), whether schools and classrooms achieve a level of diversity with respect to lived experiences is an important discussion for another forum. Rather, my position in this chapter is merely that diversity, as a concept, is quite congruent with the ideas we have explored thus far.

Similarly, equity (the "E" in DEI), is consistent with virtuous pursuits in the mathematics classroom. In fact, in our examination of justice and community, equity emerged as a key term. In the math classroom, equity refers to providing support and enrichment to students as they need it. French utopian philosopher, Étienne-Gabriel Morelly, described this to this as an equalizing according to needs (Mainali, 2025). On this point, Su (2020) writes, "justice means treating people equitably – giving each person what they are due. It means speaking up for the powerless, who are most likely to be treated unjustly" (p. 149). As teachers, I would argue that one of our fundamental responsibilities is identify needs for support and enrichment among our students and then address them such that everyone in our math classroom has what is needed to realize success. So often, this means varying our supports and practices to meet individual needs, or as it has been called for decades, *differentiation*. When we talk of equity in the math classroom, at core, we are talking about the humane (and effective) approach of meeting individual students' needs such that they can be engaged members of our math community. From my perspective, this is another element of DEI, that when thoughtfully considered, is somewhat inarguable as a key element for virtuous pursuits and human flourishing.

Lastly, the "I" in DEI stands for inclusion. In simplest terms, this refers to feeling a part of the group – or, included. Returning to our investigation of community, Su (2020) writes, "That feeling – *I don't belong* – can be quite crippling. And this is where community is terribly important: for us to feel belonging" (p. 188). As humans, we are very social beings, and that sense of belonging, of being included, is so very important to experience fulfillment. In the math classroom, inclusion describes the extent to which we work to create a community where everyone feels like they belong. No matter one's lived experience, family income, race, gender, ethnicity, religion, physical or cognitive ability, there is a place for you in this math classroom, and you are a valuable part of our group. That's the ethos of inclusion and creating classroom environments where students feel a part of, where they experience real community, open opportunities for meaning, play, perseverance, and so forth that would not be present otherwise.

Certainly, scholars in this area could provide far more elaborate descriptions of DEI elements, histories, and rationales for importance, my hope is that this rudimentary examination of each element illustrates how DEI is wholly consistent with our aims of teaching math for human flourishing and virtuous pursuits. Indeed, if we step back from the politicization of the terminology just a bit, embracing difference (diversity), providing individualized support (equity) and fostering belonging (inclusion) in our classrooms is, by my estimation, an uncontroversial approach to teaching and learning. I sincerely hope I don't alienate some readers, but DEI is deeply embedded in the virtues explored in this book and is part and parcel of teaching elementary mathematics for human flourishing. To the extent that we make DEI a big deal, I would suggest we do so by thinking about how to expand our embrace of difference as a strength, our provision of individualized supports, and our creation of a math community where all feel included.

Artificial Intelligence and Authentic Human Beings

Another topic that I encounter with increasing frequency is artificial intelligence, or AI. First off, I need to go on record that I am not a technology scholar and that my sense of how emerging tools and technologies will shape the future is little more than that of an enthusiastic layperson. However, as a mathematics educator who regularly interacts with elementary teachers, I wonder a great deal about how this technology is currently influencing our teaching and learning as well as what may be coming in the months and years ahead.

AI is a pretty broad category that could, conceivably range from a chatbot on your mobile device (e.g., Kindroid) to the way that smart devices (i.e., appliances and thermostats) learn and adapt to your inputs. For this discussion, let's focus on generative AI tools that rely on large language models (basically the entirety of the internet) to generate content (e.g., text, images, and video) based on user prompting. Perhaps you have noticed, in recent Google searching, AI-generated summaries (referred to as an AI Overview) that aim to capture essential elements of a searched-for topic. For example, a Google search for "beat poetry", yields an AI-generated summary of key elements of beat poetry akin to a brief Wikipedia entry in addition to just returning websites. Of course, this barely scratches the surface of current AI capabilities. ChatGPT 5 (via the current individual subscriber version of this particular AI tool), is highly generative and can respond to very detailed prompts, create images, music, and even video. For example, I asked ChatGPT 5 to create a poem about teaching math for justice and community in the elementary classroom in the style of Emily Dickenson and create an accompanying image. Given the current ambiguities around copyright and ownership of AI-generated outputs, I am unable to share the results, but you can easily recreate this little experiment on your own.

The first time I experimented, in this way, with this sort of generative AI tool, I was impressed with the sheer speed with which it could create content. The poem it created was a plausible facsimile of Dickenson's style. Likely, it would take me a good while to create a similar poem and draw a related image. However, if I am honest, the overall quality of these outputs was rather uninspired. The poem was fine, in so far as it goes, but there was a certain "Mad Libs" quality that renders it generic. For example, one generated line, "The chalk returns to dust, we keep the brighter equation of equity and trust" struck me as a pedestrian approach to combining certain buzzwords. Similarly, the generated image (a cartoon) was not very sophisticated at all. The picture, seemingly stylized to reflect the 1950s, was that of a teacher leading a small groups of students through math tasks written on a blackboard. The tasks on the board seem tenuously related to one another (5 + 1; 9 × 7), the students' representation even more so, and my overall impression was that it was not much more advanced than contemporary clip art. Altogether, my sense of these outputs is one of a fairly rote performance. Oddly enough, this gives me a lot of hope and optimism for our work of math teaching and learning.

Before we proceed further, if you haven't spent much time with generative AI tools such as ChatGPT 5 or Anthropic's "Claude" tool, I highly encourage you to do so. Take a few minutes, right now, to open up a browser, navigate to a generative AI site (there are free versions available), and just pose a few prompts. Get wild with your asks, and the more creative the better. Some ideas to get you started might include: "Using my social media presence, write a song about me in the style of Taylor Swift" Or "Generate five different pictures of sea otters riding bicycles in the style of Andy Warhol". Also, and this is key, follow up with a guiding prompt to see if AI can optimize its output in a way that's most desirable for you. For example, "Using my social media footprint, try to make the song about me more similar to Taylor's 1989 album" or "create a new set of five images to reflect a deep friendship amongst the otters". You might also go even deeper and more practical, "crawl through my district and school websites and create a detailed report of strengths and weaknesses regarding math instruction in my area". You get the idea.

Whatever your prompt, take a moment and jot down some notes about how well the AI tool performed. Were you impressed? If so, in what why? Is there any aspect to this output that you find less impressive? How would you describe this sense of shortfall?

____

____

____

____

____

If you took the time to play with a particular AI too, I suspect we might share similar feelings. These tools can take extremely specific asks and produce a variety of responses and outputs in a manner that is quite impressive. However, my guess is that there is also something lacking in these outputs in the same way that my prompts about math poetry and associated imagery left me wanting.

As an AI optimist, I believe there are many ways that these tools can augment math teaching and learning (and education more broadly) to help us do important work. For example, I am hopeful that the next generation of AI-powered wearable technology will provide translation capabilities such that teachers could more effectively communicate with students who are emerging English language learners. At this very moment, AI is capable of analyzing rich inputs with a goal of identifying hidden trends and areas for growth – [Sample Prompt] "analyze my lesson plans and progress monitoring outcomes for math over the last two weeks and generate ten suggestions for improving my math teaching". *Note, we want to be careful about imputing student level data into AI systems and check your school and district policies before uploading any scores our assessment information.* AI is also quite useful for optimizing particular mathematical tasks we might pose to students – [Sample Prompt] "generate five different math tasks involving adding unit fractions that would promote rich discussion among students". Again, while I am unable to share, directly, the outputs of this prompt, I will attempt to provide some description of my impressions. While the generated tasks were not quite what I had in mind, it was easy for me to see how generative AI tools might provide a range of ideas upon which we could build and tinker. For example, one suggestion was to task students with designing a "fair game" involving a spinner. They would begin with a sample spinner with regions reflecting unit fractions (e.g., 1/2, 1/3, 1/6, etc.). Each player spins twice and adds the fractions and the player with the highest total wins. More interestingly, students are tasked with making a determination about whether or not the game is fair and, if not deemed fair, how the spinner might be redesigned. Again, as generated, this task might not be an ideal fit for my particular classroom, but it might provide a jumping off point to propel my own pedagogical thinking and design work.

While these sorts of generative tools can help us initiate design efforts, the ideas we have traversed in this book are much bigger and ground us in far more human spaces. It is difficult to imagine, for example, how AI tools might help us foster love, pursue truth, or appreciate beauty in our math classrooms. No doubt, AI might generate some task ideas or possible discussion prompts, but the real work of connecting math teaching and learning to these virtues, I believe wholeheartedly, requires human beings. Specifically, it is our task as teachers to *rehumanize* (Gutiérrez, 2018) the mathematical experience for our students. This is the foundation for my optimism. I concede that generative AI tools are developing with increasing rapidity, and the outputs five years from now may be wildly different than those we are considering presently. However, I remain skeptical that AI could ever engage with the virtues of this book in the ways that we, as human beings, can. Returning to my optimism, to the extent that we may be able to off-load the more mundane analytic and logistical-oriented tasks to AI tools, we create space to focus on humanizing the experience via virtuous pursuits that result in deeper fulfillment. There is a path, interestingly enough, where the increasing deployment of AI tools in educational spaces leads to a more humane and virtuous experience among our students. However, this requires us, as teachers, to claim that right. We must grasp that opportunity afforded to us and help those around us recognize the value of connecting these deeply human pursuits to the math experiences of our students. Nothing does "human" quite like a human being.

From Me to You

Since I began work on this book, I have thought a lot about this conclusion. Returning to the guidance I received long ago to begin with the end in mind, I grappled with the key ideas with which I hoped readers might take from this work. Given the extraordinary richness of the topics at hand, there are so many places we could go. However, if I distilled my hopes for where this book leaves us, I arrive at two things – *joy* and *advocacy*. Think of these ideas as parting takeaways and a call to action.

Joy is yet another virtuous pursuit that can deeply inflect our lives and those of our students. While the experience of joy is often quite personal and contextual, arguably the pursuit and experience of virtues we have examined in this book may unlock joy that we feel when engaging in some deeply human activity or experience. That is, seeking meaning, appreciate beauty, working for justice, and so forth, propel us in a joyful direction – toward a life fulfilled. A rewarding mathematical program intentionally creates space for joy in learning designs and celebrates such joy when it occurs. Importantly, this experience of joy is not only for students, but teachers as well. Su (2017) writes,

> by joy, I refer to the wonder or awe or delight in the beauty of the created order. By transcendence, I mean the ability to embrace mystery of it all. There's a transcendent joy in experiencing the beauty of mathematics. Mathematics makes the mind its playground. We play with patterns, and within the structure of certain axioms, we exercise freedom in exploring their consequences, joyful at any truths we find. We even have a whole area of mathematics known as recreational mathematics.

There is so much opportunity to find joy in our work, but it is so easy to lose sight of this in the highly scheduled pace of a school day. Joy is so very important, though. Joy grounds us in the moment and also fosters optimism about what may come. Joy breeds appreciation, gratitude, kindness, and empathy. And joy, is often just pretty darn thrilling in its own right.

The reason that I wanted joy to be one of the topics with which we part is that, at this moment, I am worried about my profession. As I mentioned in the introduction, the Learning Policy Institute has been tracking a teacher shortage for some time. According to their most updated reporting, approximately *one out of every eight* teaching positions is either unoccupied or occupied by an individual who lacks appropriate training and certification (Learning Policy Institute, 2025). We are having a really hard time attracting young people into the teaching profession. Of course, the reasons are varied, and persistent perceptions of lagging compensation definitely plays a part. However, I would argue that young people contemplating a future in teaching likely do not see schools as a place of joy. Imagine what might happen if we were to seek, embrace, and celebrate joy in our math teaching. What if we found joy in our work and that joy spread contagiously to those around us as it often does? If students began seeing joyful teachers and finding joy themselves in a math class, of all places, and I wonder how that might shape their thoughts about doing this work of teaching. In a landscape where state and local budgets are almost always under pressure, recasting math classrooms (and schools) as places of joy can only help. It all begins with us, and my hope is that our time spent with these ideas helps you unlock joy and share it with those around you.

Lastly, I want to leave you with a call for advocacy. If you found this book to be compelling, then a good next step is to talk about math teaching and these virtues with those around you. The discipline of mathematics education continues to be contested space. That is, we, as a society, struggle to agree on what learning mathematics entails, why it's useful to learn, how it should be learned, and so on. Some argue that mathematical learning experiences should focus on factual knowledge and internalized procedures that are developed by direct instruction. Others, like me, argue that math teaching and learning can be a space of healthy struggle, perseverance, exploration, and play. That's not to say that internalized knowledge and procedures don't matter. They do, for certain. Knowing my multiplication tables or how the commutative property works, somewhat instinctively, frees my mind up to do deeper and more fulfilling work. The key is that last part, though. Math involves building some core knowledge and skills, but it is also so much more. Math is a doorway through which we might lead our students into a new world where they experience beauty, find meaning, feel empowered, and ultimately see themselves as capable young mathematicians. Again, if this vision for math teaching and learning resonates with you, talk about it with the people in your orbit. Ask them about their own math experiences? Are there places, in the lives of those around you, where connection to virtue and a more rehumanized approach to math might have changed things for the better? What about the leaders in your school, district, and community? What do they think math education should entail? What are their goals for young people with respect to math? I would argue that the ideas of this book are not necessarily contradictory to many of the prevailing ideas about how we should teach math but rather expand upon them. From my perspective, it's not an *either/or*, but rather a *yes/and* proposition. My goal, in these conversations, is to help those around me think about how we can dream bigger in our classrooms, help our students experience more, and guide them toward being fulfilled individuals ready to bring their mathematical minds to bear on the world around them. If you find this vision inspiring, then as they say in my neck of the woods, go tell it on the mountain.

In closing, to all of you readers, I want to offer my most sincere thanks and gratitude. If you are a teacher or involved in education in some capacity, I bid my deepest appreciation and cheer for your commitment to the betterment of young people. For those in other professional spaces that, for whatever reason, encountered this book, thank you for spending time with me. No matter your background, I hope these ideas bring you closer to fulfillment and maybe brought you a little dose of mathematical fun and insight along the way. I find these sorts of ideas remarkably exciting, and if you do too, I welcome you to reach out. In the meantime, keep persevering and seeking virtuous connections wherever you may find yourself, and be mindful of your students and the young people around you. They are ever watchful and look closely to us to see how we spend our time, where we direct our energy, and exactly what it is we value. Let us share with them the good stuff.

References

Charles, C. Z. (2003). The dynamics of racial residential segregation. *Annual Review of Sociology*, *29*(1), 167–207.

Denton, N. A. (1995). The persistence of segregation: Links between residential segregation and school segregation. *Minnesota Law Review*, *80*, 795.

Foss, K. A. (2020, October 5). Remote learning isn't new: Radio instruction in the 1937 polio epidemic. *The Conversation*. Retrieved from https://theconversation.com/remote-learning-isnt-new-radio-instruction-in-the-1937-polio-epidemic-143797

Gutiérrez, R. (2018). The need to rehumanize mathematics. In I. Goffney, R. Gutiérrez, & M. Boston (Eds.), *Rehumanizing mathematics for*

Black, Indigenous, and Latinx students (pp. 1–10). Reston, VA: National Council of Teachers of Mathematics.

Learning Policy Institute. (2025). An overview of teacher shortages: 2025 [Fact sheet]. *Learning Policy Institute.* https://learningpolicyinstitute.org/products/fact-sheets

Mainali, G. P. (2025). Utopian socialist thought as the foundation of Marxist perspective. *Interdisciplinary Journal of Management and Social Sciences*, *6*(1), 136–142.

National Governors Association (NGA) Center for Best Practices & Council of Chief State School Officers (CCSSO). (2010). *Common core state standards for mathematics.* Washington, DC: Authors.

Su, F. (2017). Mathematics for human flourishing [Blog post]. *The Mathematical Yawp.* https://mathyawp.wordpress.com/2017/01/08/mathematics-for-human-flourishing/

Su, F. (2020). *Mathematics for human flourishing*. New Haven, CT: Yale University Press.

For Product Safety Concerns and Information please contact our EU representative GPSR@taylorandfrancis.com
Taylor & Francis Verlag GmbH, Kaufingerstraße 24, 80331 München, Germany

www.ingramcontent.com/pod-product-compliance
Ingram Content Group UK Ltd.
Pitfield, Milton Keynes, MK11 3LW, UK
UKHW022119190926
475702UK00013B/1543
* 9 7 8 1 0 3 2 7 7 7 8 7 0 *